SD329
Science: Level 3

The Open Univ

Signals and perception:
the science of the senses

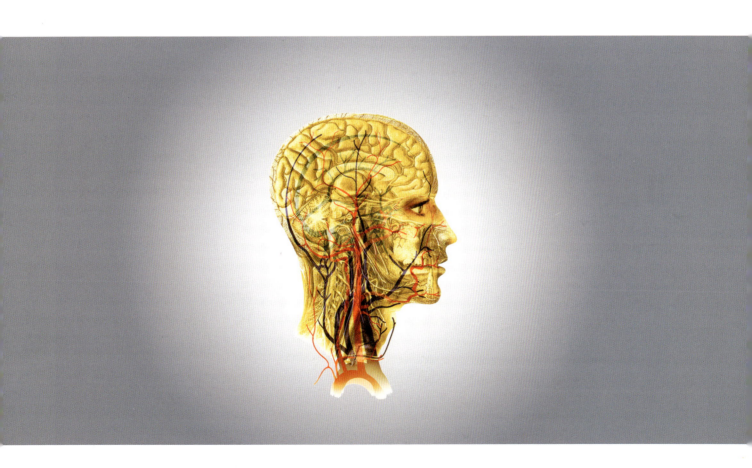

Block 4 Vision

This publication forms part of an Open University course *SD329: Signals and Perception: the science of the senses*. The complete list of texts which make up this course can be found at the back. Details of this and other Open University courses can be obtained from the Student Registration and Enquiry Service, The Open University, PO Box 197, Milton Keynes MK7 6BJ, United Kingdom: tel. +44 (0)870 333 4340, email general-enquiries@open.ac.uk

Alternatively, you may visit the Open University website at http://www.open.ac.uk where you can learn more about the wide range of courses and packs offered at all levels by The Open University.

To purchase a selection of Open University course materials visit http://www.ouw.co.uk, or contact Open University Worldwide, Michael Young Building, Walton Hall, Milton Keynes MK7 6AA, United Kingdom for a brochure. tel. +44 (0)1908 858793; fax +44 (0)1908 858787; email ouw-customer-services@open.ac.uk

The Open University
Walton Hall, Milton Keynes
MK7 6AA

First published 2002. Reprinted 2007.

Edited and designed by The Open University.

Typeset by The Open University.

Printed and bound in the United Kingdom by Latimer Trend & Company Ltd, Plymouth.

ISBN 0 7492 3577 2

1.3

The paper used in this publication contains pulp sourced from forests independently certified to the Forest Stewardship Council (FSC) principles and criteria. Chain of custody certification allows the pulp from these forests to be tracked to the end use (see www.fsc.org).

The SD329 Course Team

Course Team Chair
David Roberts

Course Manager
Yvonne Ashmore

Course Team Assistant
Margaret Careford

Authors
Mandy Dyson (Block 3)

Jim Iley (Block 6)

Heather McLannahan (Blocks 2, 4 and 5)

Michael Mortimer (Block 2)

Peter Naish (Blocks 4 and 7)

Elizabeth Parvin (Blocks 3 and 4)

David Roberts (Block 1)

Editors
Gilly Riley

Val Russell

Indexer
Jean Macqueen

OU Graphic Design
Roger Courthold

Jenny Nockles

Andrew Whitehead

CD-ROM and Website Production
Jane Bromley

Eleanor Crabb

Patrina Law

Kaye Mitchell

Brian Richardson

Gary Tucknott

Library
Judy Thomas

Picture Research
Lydia Eaton

External Course Assessors
Professor George Mather (University of Sussex)

Professor John Mellerio (University of Westminster)

Consultants
Michael Greville-Harris (Block 4, University of Birmingham)

Krish Singh (Block 2, Aston University)

BBC
Jenny Walker

Nicola Birtwhistle

Julie Laing

Jane Roberts

Reader Authors
Jonathan Ashmore (University College London)

David Baguley (Addenbrooke's Hospital, Cambridge)

Stanley Bolanowski (Syracuse University)

James Bowmaker (University College London)

Peter Cahusac (University of Stirling)

Christopher Darwin (University of Sussex)

Andrew Derrington (University of Nottingham)

Robert Fettiplace (University of Wisconsin)

David Furness (Keele University)

Michael Greville-Harris (University of Birmingham)

Carole Hackney (Keele University)

Debbie Hall (Institute of Hearing Research, Nottingham)

Anya Hurlbert (University of Newcastle upon Tyne)

Tim Jacob (University of Cardiff)

Tyler Lorig (Washington and Lee University)

Ian Lyon (Consultant)

BLOCK FOUR

VISION

Contents

1 Introduction   **5**

2 The signal   **9**

   2.1 What is light?   9

   2.2 Sources and surfaces   11

   2.3 Colour science   13

   2.4 Spatial variation of intensity   15

   2.5 Summary of Section 2   19

3 The detector: the human eye   **21**

   3.1 The structure of the eye   21

   3.2 The pupil and the iris   22

   3.3 The dioptric apparatus   23

   3.4 The retina   36

   3.5 Colour vision   43

   3.6 Movement of the eye   45

   3.7 Adaptation to different levels of illumination   50

   3.8 Visual acuity   56

4 From eye to brain   **71**

   4.1 Overview   71

   4.2 Pathways from the retina to the cortex   71

   4.3 Visual sub-modalities   76

   4.4 Putting it all together   86

   4.5 The next step   91

   4.6 Summary of Section 4   92

5 Recognizing shapes   **93**

   5.1 The beginnings of a theory   93

   5.2 Grandmother cells   94

   5.3 The timing problem   94

   5.4 Reading for meaning   95

5.5	The perception of faces	95
5.6	'Don't I know you from somewhere?'	95
5.7	Facial analysis	96
5.8	Processing in context	97
5.9	Bottom-up and top-down processing	99
5.10	The connectionist account	100
5.11	Summary of Section 5	102

Objectives for Block 4 103

Answers to questions 104

Acknowledgements 109

Glossary for Block 4 110

Index for Block 4 123

Introduction

1

This block covers the sense of vision. Many people would say that vision is the most important human sense; it is certainly the best developed, as witnessed by the size of the visual cortex in comparison with the regions of the brain devoted to the other senses. The basic structure of the eye, familiar to many, is shown in Figure 1.1.

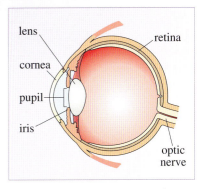

Figure 1.1   A diagram showing the basic structure of the human eye.

Consider what amazing things the normal human eye can do:

- detect light over the range of wavelengths 380–780 nm (1 nm = 10^{-9} m) (Figure 1.2);

- focus this light to form a sharp image of any object that is more than 250 mm from the eye;

- resolve detail within less than 1 mm at a distance of 250 mm. That means that most people can read <small>writing this small;</small>

- operate over a range of light levels that differ by a factor of more than 10^{12};

- distinguish colours of the spectrum that differ by only a few nm in wavelength;

- move so that any object within a horizontal angular range of at least 120° can be focused on the centre of the retina of both eyes (or more than 120° allowing for monocular vision only).

These are the properties of the eye itself, and will be covered in Sections 2 and 3 of this block.

Figure 1.2   The human eye is sensitive to wavelengths between 380 nm and 780 nm.

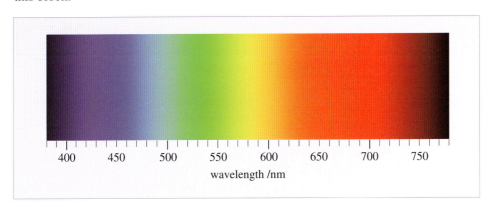

But vision is much more than just the formation of an image on the retina by the optical system. Light signals at the two retinae are converted to electrical signals and sent to the brain. How this takes place is the subject of Section 4. This conversion of light to electrical signals poses several questions. For example: what is meant by 'colour'? Neural signals are not coloured so how does the eye distinguish colours?

In Section 5 we discuss how the components of a visual scene that were discussed in Section 4 can be combined to make an object that we can recognize.

Perception usually allows us to make sense of the images we receive; on the other hand it can sometimes deceive us, something that occurs in visual illusions. In perception our brains combine signals from the eye with what might be described as 'previous experience' in order to make sense of the world around us. For example the brain can interpret shapes from lines. Look at Figure 1.3a (overleaf), a sketch composed of a few lines. You immediately *know* that it is a picture of a bicycle; a bicycle with circular wheels. But look again at the wheels: you did not *see* circular wheels, you saw ellipses as shown in Figure 1.3b.

Figure 1.3 (a) The brain has no problem interpreting this line sketch as a bicycle and 'seeing' the wheels as circular, but when isolated from the rest of the picture (b), the wheels are clearly not circular.

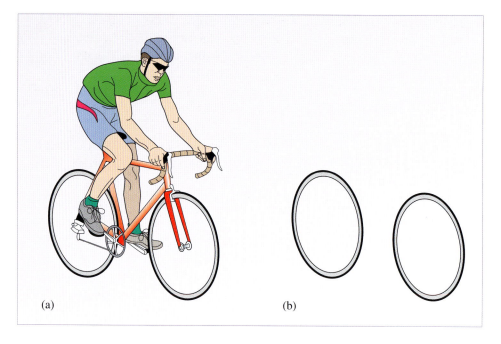

(a)   (b)

The brain has used its previous experience of bicycles, and the image that they present to the eye from different angles, to deduce that this picture is of a bicycle. This is just one example of how the brain can interpret shapes from lines.

The brain can also perceive depth. There are several ways it can do this. One is by receiving two slightly different images that are laterally displaced due to the position of the eyes – binocular vision. However it can also be done from other visual cues. Look at Figure 1.4. This is a flat image, but cues about perspective, occlusion (one object hiding another), shading and the sizes of familiar objects all combine to tell you that this is a real three-dimensional scene. Note that again you need experience (i.e. learning needs to have occurred) before you can judge that this is a three-dimensional scene.

Figure 1.4 A painting of the Grand Canal in Venice by Canaletto. This is a 2D image but is seen as a 3D scene. This perception of depth has nothing to do with having two eyes. In fact the depth effects can be greater if you close one eye.

Seeing movement is essential for survival – for people as well as for animals. And the human eye is very sensitive to motion. The edges of the field of vision are sensitive only to movement. If an object moves at the side of your visual field you can see movement even if you cannot determine what the object is. When you see movement like this 'out of the corner of your eye' it normally stimulates a reflex that causes you to turn your head to look at the moving object.

Movement can be perceived either when the object is moving relative to the eye or when the eye follows an object and keeps it steady on the centre of the field of view. Does the latter perception of motion depend on being able to see the background moving on the retina?

If you were to ask someone to wave a dim pencil torch about in a dark room, would you see it moving? In this case you could not see the background so how would your brain know that the torch is moving?

In contrast to the complications that obviously exist in the study of motion, you might think that perhaps the perception of colour is more straightforward. After all, surely we can link each colour to a particular wavelength? Try the exercise described in Figure 1.5.

Figure 1.5   Stare at the cross in the centre of the blue patch for 30 seconds. Then transfer your gaze to the centre of the yellow patch. The yellow patch now appears to have a central area of a brighter shade of yellow.

The effect you just observed has its origin in the retina. However, there are also other effects related to perception. One of the most important of these is colour constancy. Figure 1.6 shows the effect whereby the same shade of grey can look different in different circumstances. But most of the time this is not a problem – we see the colour of a red shirt or a green leaf as constant even if the conditions of illumination change (Figure 1.7 overleaf).

What about reading, a higher order skill that you must possess in order to be reading this block? To complete this introduction you might like to try the exercise described in Figure 1.8 (overleaf), which demonstrates the Stroop effect. You will meet this again in Section 5.

Phenomena such as these serve to illustrate the complexity of vision and the range of disciplines needed to study it. Physics is used to study the optical performance of the eye, chemistry to look at the processes occurring in the retina, neurobiology to examine the transfer of the signals to the brain and psychology to consider perception. This block necessarily covers a very wide range of material because it is only by pulling together the knowledge obtained from all these disciplines that we can begin to have any understanding of the complex process called vision.

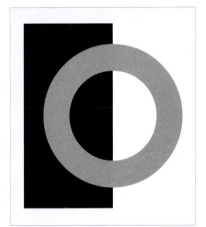

Figure 1.6   The part of the grey ring viewed against the black background appears lighter than the rest of the ring viewed against the white background. The effect is enhanced if you place a pen vertically along the line between black and white.

(a)   (b)

Figure 1.7   The same collection of fruit is illuminated by (a) tungsten, and (b) fluorescent light. The images are now entities in themselves, so they do look different. However, viewing the fruit in the context of different lights would give a more constant perception.

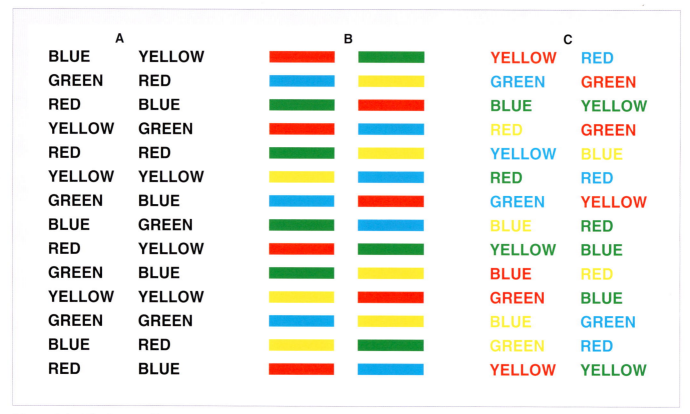

Figure 1.8   The Stroop effect. The times required to read the list of colour names in column A or to name the colour of each patch in column B are significantly less than the time needed to name the colours in the list of coloured words in column C.

The signal

2.1 What is light?

Before we look at the eye and how it works we need to ask what the signal is that is being detected by the eye. The glib answer to that question is, of course, 'light'. But what do we mean by 'light'? The word means different things to different people. We all know that we need light to see and most of us can distinguish different 'colours' of light. But what are these different colours?

An artist will talk about the 'quality' of the light, and a cricket player will even talk about 'bad light'. If we want to be scientific about it we would say that they are making a comment on the overall intensity, the relative intensity of the different wavelengths, and the direction of the light.

Scientists will often talk about 'light' when they actually mean the whole of the electromagnetic spectrum (Block 1). But the human eye is only sensitive to a very small range of wavelengths between about 380 nm and 780 nm*. If we wish to draw a comparison with hearing, this is roughly an octave! Perhaps the first question we should therefore ask about vision is why is it that sight is such an important sense when it detects such a small part of the spectrum?

One answer to that question lies in the properties of sunlight. The surface temperature of the Sun is approximately 6000 K. It emits radiation, known as **black body radiation**, which, for a body with a surface temperature of 6000 K, peaks in intensity at about 500 nm. So perhaps it is not surprising that our eyes (and the eyes of most animals) have evolved to be sensitive to approximately the range of wavelengths emitted most strongly by the Sun. Some other stars have higher or lower surface temperatures and the radiation emitted peaks in intensity at shorter or longer wavelengths (Figure 2.1 overleaf). One might speculate that life near those stars might evolve to be sensitive to a different range of wavelengths.

A second answer is that this narrow range of wavelengths provides information that is at an appropriate spatial scale to allow us to interact with the world. It tells us about useful properties of the world such as the shape and surface consistency of objects. Shorter wavelengths, for example, could potentially provide us with different information about the microworld of molecules and atoms – but such information would be of very little use in guiding our everyday activities.

You saw in Block 1 that light can be considered as an **electromagnetic wave**. As the wave is transmitted there is a periodic variation in both the electric and magnetic fields. This variation of the fields does not require any medium in which to travel, so an electromagnetic wave can pass through a vacuum as well as through air, water, glass or any other substance. You may recall that the speed of the wave in a vacuum (c) is always exactly the same and has a fixed value very close to $3.0 \times 10^8 \, \text{m s}^{-1}$. The speed in air has approximately the same value as in a vacuum; in all other media the speed is reduced. As with all waves the speed is the multiple of the wavelength and the frequency so that:

$$c = f\lambda \qquad (2.1)$$

where c is the speed, λ is wavelength and f is frequency.

* Strictly speaking this is the maximum range. The sensitivity at the ends of the range is very low, and for most diagrams we shall take the range as 400–700 nm.

Figure 2.1   The radiation emitted by the Sun (orange line) and hotter (9000 K) and cooler (4000 K) stars. Intensity is plotted against wavelength.

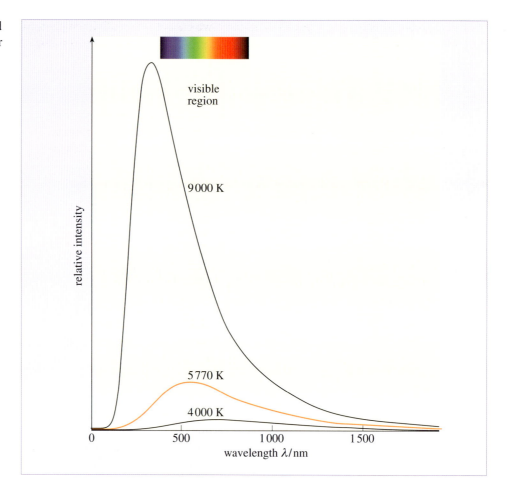

The understanding that visible light was a form of electromagnetic radiation arose out of the work of James Clerk Maxwell in the second half of the nineteenth century. The first half of the twentieth century saw the development of a new branch of physics – quantum theory. Experiments such as those investigating the photoelectric effect showed that, while the wave model of light works very satisfactorily when considering the propagation of light, it cannot explain the interaction of light with matter. The new quantum theory required an alternative model according to which light is treated as packets of energy, **photons**, which interact individually with atoms. Each photon has an energy that depends on the frequency of the light and that is given by the expression:

$$E = hf \tag{2.2}$$

where E is the energy, f is the frequency and h is Planck's constant (6.626×10^{-34} J s).

○   Which has more energy – a blue photon or a red photon?

●   Blue light has a larger value of f than red light so a blue photon has more energy.

The modern view is that light exhibits wave–particle duality. Under certain circumstances, usually those relating to the propagation of light, we must use the wave model to provide an explanation of what happens. On the other hand, when light interacts with matter, as it does at the retina, the photon model, which treats light as a stream of particles, is needed. Although we cannot 'see' individual photons (it requires several photons for us to be aware of a short flash), the receptors in the retina can be stimulated by just one.

2.2 Sources and surfaces

The sensation that we describe as 'colour' depends on the wavelength, or mix of wavelengths, that reaches the eye. There are a few sources that emit only one wavelength (these are known as **monochromatic**, meaning 'one colour'). A familiar example is a sodium streetlight from which the emitted light is almost entirely of one wavelength, 589 nm. And there are some sources that produce a very limited range of wavelengths and therefore look coloured (e.g. traffic lights). But most useful sources of light emit a wide range of wavelengths. Figure 2.2 (overleaf) shows the **spectral distribution** of the light emitted by a few common sources represented mathematically by the function $E(\lambda)$. Such sources are known as '**white light**' sources because the effect of mixing all the wavelengths is to give the sensation of white.

For sources such as these the light we see is emitted directly from the source. However most of the light reaching our eyes is not coming directly from a light source such as the Sun – it has been reflected off an object. So why does an object appear coloured if the light from the source includes a wide range of visible wavelengths?

Objects appear coloured because they reflect some wavelengths and absorb others. Take the simple example of a tomato sitting in a bowl in your kitchen. White light, perhaps from the Sun, or from a tungsten light in your home, is striking the tomato. But only light giving the sensation of an orange-red colour is being reflected back to your eye – the blues and greens are being absorbed and only reds and some yellows are reflected.

○ Sketch the form of the spectral distribution curve for the light reflected from the tomato.

● It would contain very little light from the blue end of the spectrum but plenty from the yellow and red regions. Your sketch should look something like the red line in Figure 2.3 (overleaf).

So the amount of light reflected from an object at a particular wavelength depends both on the reflecting properties of the object (i.e. whether it is a tomato that reflects red light or an orange that reflects orange light) and on the incident light. This point is illustrated in Figure 2.4 (overleaf) where the light reflected from the tomato is shown for a different light source.

If we describe the reflected light by the function $I(\lambda)$ (also a function of wavelength) then we can work out the **spectral reflectance** of the tomato, $S(\lambda)$. This function is something that depends only on the composition of the tomato and not on the lighting conditions. The three functions are related by the equation:

$$I(\lambda) = S(\lambda) \times E(\lambda) \qquad\qquad (2.3)$$

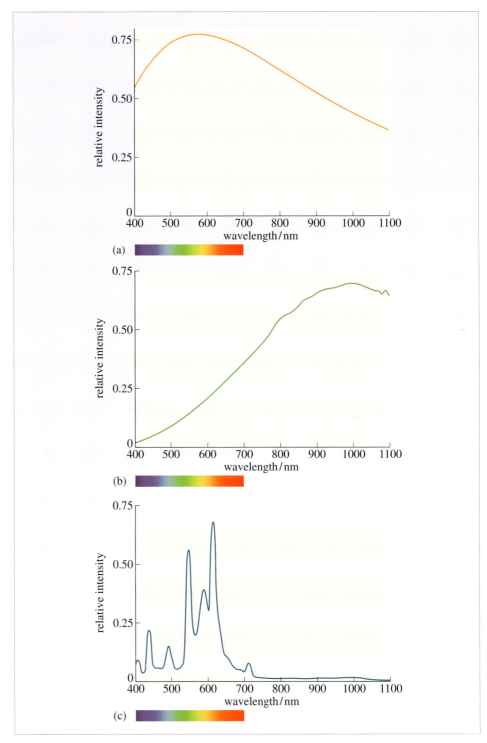

Figure 2.2   The spectral distribution of light from (a) the Sun, (b) a tungsten lamp, and (c) a fluorescent lamp. The curve is represented mathematically by the function $E(\lambda)$ where E represents the amount of light emitted at wavelength λ. Note that the tungsten light emits more radiation in the infrared region which is of course not visible.

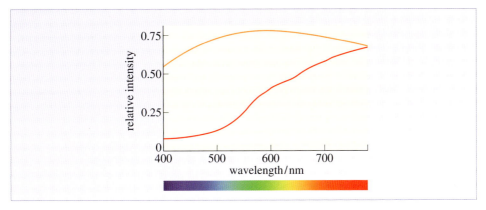

Figure 2.3 Reflection of light from a tomato. The spectral distribution for the source (sunlight) is shown by the orange line, and the reflected light by the red line.

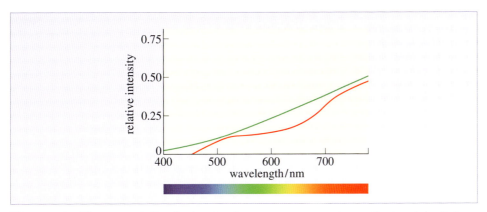

Figure 2.4 The spectral distribution for the tungsten source (green line) and the light reflected from the tomato (red line).

Remember that in each case you need to multiply the values of each function at a particular wavelength. Figure 2.5 (overleaf) shows how the spectral distribution, $E(\lambda)$, for the source (sunlight) multiplied by the spectral reflectance of the tomato, $S(\lambda)$, gives the spectral distribution of the light entering the eye, $I(\lambda)$.

2.3 Colour science

Block 1 introduced the ideas of additive and subtractive colour mixing. You will remember that all colours of light can be achieved by mixing the three **additive primary colours**, red, green and blue. The fact that there are three primary colours actually tells us something important about the properties of the human eye, which will be discussed in Section 3. None the less this is an appropriate point to discuss various aspects of colour science.

You should now read Chapter 8 of the Reader, *Colour science* by Stephen Westland. In this he introduces the ideas of **hue**, **value** and **chroma** and looks at the usefulness of the **CIE chromaticity diagram** in defining colours. You should aim to get a general understanding of the ideas of colour science rather than being concerned with mastering the mathematics.

Figure 2.5 The spectral distribution, $E(\lambda)$, for the source (sunlight) multiplied by the spectral reflectance of the tomato, $S(\lambda)$, to give the spectral distribution of the light entering the eye, $I(\lambda)$. The tomato reflects the red and yellow light but absorbs the blue and green light.

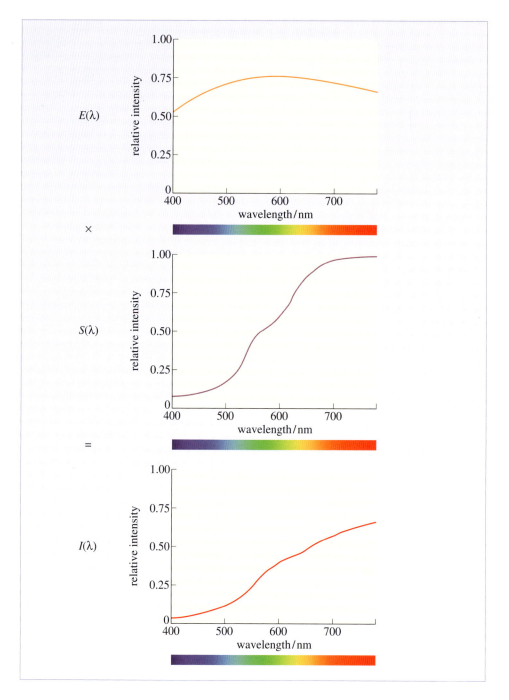

Question 2.1

There is only one word for hue but more than one word for value and chroma in the chapter. What are the synonyms?

Activity 2.1 Colour science

At this point you should undertake the CD-ROM activities related to colour science. Further instructions are given in the Block 4 *Study File*.

2.4 Spatial variation of intensity

When we look at a scene, the image formed on the retina varies in intensity from place to place. This variation in intensity (often known as contrast) is one of the major ways in which we pick up information about the scene. So one of the major functions of the eye is to distinguish *variations* in intensity.

Sometimes we are very much aware of a pattern in these variations, as for example when we look at the black and white stripes of a Newcastle United football shirt; at other times we are not aware of any regularity in the image at all. However, it is possible to analyse all images in terms of their **spatial frequencies**.

The idea of spatial frequency will probably be easier to understand if we start off by thinking about **temporal frequencies**. In Block 3 you came across the idea that a periodic wave, such as the sound wave produced by a musical instrument, can be shown to be the sum of several different waves with different frequencies and usually with different amplitudes and phases. The process of describing the sound by giving the amplitudes and phases of its component frequencies is known as Fourier analysis. The result is known as a Fourier series if the waveform is periodic, or, more generally, as a Fourier transform.

Temporal frequencies are given in hertz, or cycles per second. They are very useful when considering the auditory stimuli to the ear, but in vision one is much more concerned with variations in *space*. Figure 2.6 shows a waveform that is periodic in one-dimensional space and has been broken down into its spatial frequencies.* The mathematics is just the same as for temporal frequencies – the only difference is that the frequency is now expressed in units of 1/distance rather than 1/time. The symbol q and the units 'cycles mm^{-1}' are often used for spatial frequencies. However, when considering the eye, it is easier to use 'cycles $degree^{-1}$' as this makes allowance for the eye–object distance. We shall use cycles $degree^{-1}$ from here on. They are measured as shown in Figure 2.7 (overleaf).

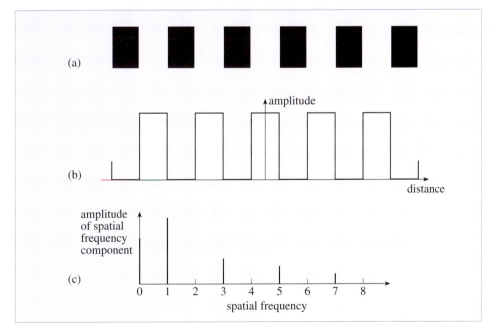

(a)

(b)

(c)

Figure 2.6   (a) A simple black and white pattern. This is what you see. (b) A graph representing the spatial distribution of the pattern with equally spaced white (amplitude 1) and black (0) regions. (c) The representation of this pattern in terms of the amplitudes of the spatial frequencies present. The horizontal axis is in units of q divided by the fundamental spatial frequency q_1. (In this example spatial frequencies are measured in units of cycles $distance^{-1}$ but from here onwards they will be measured in cycles $degree^{-1}$.)

* Be very careful not to confuse these ideas of spatial frequency with the (temporal) frequency of the light which determines the colour.

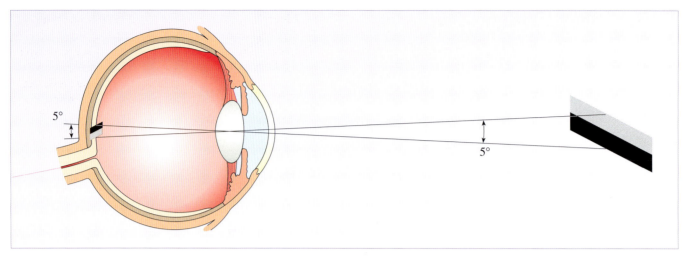

Figure 2.7   Spatial frequencies are best measured by working out the angle subtended at the eye by one dark/light cycle. This illustration shows 1 cycle in 5 degrees, equivalent to 0.2 cycles degree^{-1}.

Spatial frequency can be a difficult concept to understand, but you should appreciate that the overall shape of an object is conveyed by the low spatial frequencies and the detail is contained in the high spatial frequencies. If the high spatial frequencies are lost (because of defects in the observer's eye or the limitations of the printing process for example) then the detail of the image is lost. Figure 2.8 illustrates increasing spatial frequency, and Figure 2.9 is an interesting example of how high spatial frequencies can be lost (in this case by the image being reflected from the surface of the water).

Figure 2.8   Black/white line pairs of increasing spatial frequency. On the left the spatial frequency is low. As you go from left to right the spatial frequency increases.

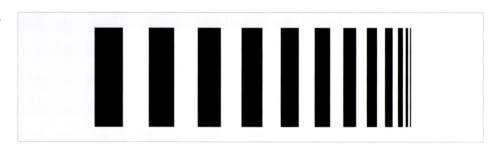

Of course, images are two-dimensional rather than one-dimensional. However, this does not present a problem – we can break down the different dimensions and carry out a Fourier transform on each of the two possible perpendicular directions.

○   Patterns such as that shown in Figure 2.10 are commonly used for research into vision and are known as gratings. Sketch the graph of intensity versus distance for the grating in Figure 2.10.

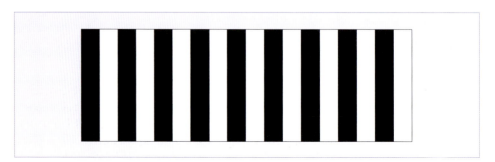

Figure 2.10   An example of a grating.

Figure 2.9   This photo of a waterside scene is an excellent example of the effect of losing high spatial frequencies. The camera used to take this photograph has been able to record high spatial frequencies in the upper part of the photo. However the image reflected in the water has lost the high spatial frequencies because the surface of the water is not completely smooth. The result is a more blurred, although still attractive, image of the buildings.

● The graph of the intensity versus distance should look like Figure 2.11.

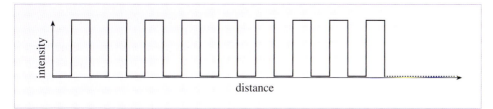

Figure 2.11   Intensity versus distance for the grating in Figure 2.10.

○ If the grating is viewed from a distance of 1 m, what is the highest spatial frequency present at the eye?

● Measurement of the grating shows that one cycle of the largest spatial frequency in the pattern (the dominant one) covers a distance of 1 cm. The angle subtended at the eye can be worked out by saying that the tangent of the angle is 1 cm/1 m = 1/100, giving an angle of approximately 0.6 degrees. So we have one cycle for every 0.6 degrees, and the number of cycles per degree is 1/0.6 = 1.67.

The highest spatial frequency present is 1.67 cycles per degree. As a rule of thumb, 1 cm subtends 1 degree of visual angle at a distance of 57 cm and, for ease of calculation, many visual experiments are conducted at a multiple of this viewing distance.

Spatial frequencies are very important in the study of all imaging processes, not just vision. In particular it is important to look at the way in which any imaging system, be it the eye or a camera, transmits different spatial frequencies. We shall return to this idea in Section 3 when we look at visual acuity.

Activity 2.2 Spatial frequencies

To reinforce these ideas about spatial frequencies and gratings you should now undertake these CD-ROM activities. Further details are given in the Block 4 *Study File*.

Question 2.2

The visible spectrum covers the wavelength range 380–780 nm.

(a) Which end of the range corresponds to violet?

(b) What is the frequency of red light that is just visible?

(c) Is this higher or lower than the frequency of violet light?

Question 2.3

What would be the effect on our lives if our eyes were only sensitive to radiation in the range 700–10 000 nm?

Question 2.4

Figure 2.12 shows the spectral reflectance $S(\lambda)$ versus wavelength for an object.

(a) What colour would it appear to the human eye if viewed in sunlight?

(b) What colour would it appear to be if viewed in the light from a sodium streetlight?

Question 2.5

What colour would be obtained by mixing red and green light? What colour would be obtained by mixing the same colours in paint? Why?

Question 2.6

Decide whether each of the following statements about the CIE chromaticity diagram are true or false.

(a) Spectral colours lie along the lower horizontal boundary of the diagram.

(b) The two-dimensional diagram in Chapter 8 of the Reader, *Colour science*, (Figure 9) shows hue and chroma but not value.

(c) Subtractive mixtures of two spectral colours lie on a straight line joining those colours.

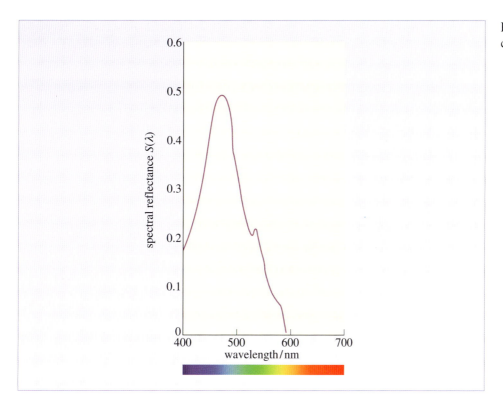

Figure 2.12 Spectral reflectance curve for the object in Question 2.4.

Question 2.7

What examples of everyday objects can you think of that would give rise to high or low spatial frequencies on the retina when viewed at a distance of a few metres? Are the patterns one- or two-dimensional?

2.5 Summary of Section 2

The eye detects electromagnetic radiation in the range of wavelengths 380–780 nm, known as the visible region. These waves have the usual relationship between speed, frequency and wavelength, $c = f\lambda$ and their speed in air is $3 \times 10^8 \, \text{m s}^{-1}$. They do not require any medium for propagation.

When light interacts with atoms it is necessary to use the photon model. According to this, light can be considered to be a stream of particles, each one with energy hf.

Very few sources of light consist of only one wavelength. For most sources the spectral distribution $E(\lambda)$ is a smooth curve.

Objects appear to have different colours depending on which wavelengths are reflected or absorbed. The light reflected by an object is a function of both the incident light $E(\lambda)$ and the spectral reflectance $S(\lambda)$. The two are linked by the equation $I(\lambda) = S(\lambda) \times E(\lambda)$, where $I(\lambda)$ is the reflected light.

For human vision the primary colours of light are red, green and blue. Light from different sources can be mixed additively. Printing systems on the other hand rely on subtractive mixing for which the primary colours are cyan, magenta and yellow.

The three properties of colour are hue, value and chroma. The three attributes can be incorporated into a three-dimensional colour system. The CIE colour specification system allows colour to be communicated, measured and controlled.

Variations in intensity can be analysed in terms of the spatial frequencies present. The general form of a pattern is given by the low spatial frequencies, high spatial frequencies give the detail. In vision it is most convenient to measure spatial frequencies in units of cycles per degree.

The detector: the human eye

Having looked at the signal that reaches the eye let us now turn to look at the eye itself. We shall start by following the path of the light from the front of the eye to the retina and looking at the conversion of photons into electrical signals in the retina. How these signals give rise to colour vision will be covered in Section 3.5. We shall then consider three topics that relate to the whole of the eye: movement (Section 3.6), adaptation to different levels of illumination (Section 3.7) and visual acuity (Section 3.8). The pathways from the eye to the brain and the way in which the brain receives and interprets those signals will be considered in Section 4.

3.1 The structure of the eye

Figure 3.1 shows a cross section of the eye. We can usefully divide this into several different systems:

- The **dioptric apparatus**, consisting of the **cornea** and the **lens**, the function of which is to form an image on the retina. In considering image formation we must also include the **iris**, which behaves as a diaphragm, regulating the amount of light entering the eye through the **pupil**, and the **ciliary muscle** and **zonular fibres** which alter the curvature of the lens.
- The muscles which allow the eyes to be moved and the eyelid to cover the eye.
- The **retina**, the purpose of which is to convert the image into electrical signals and to begin the processing of these signals. The receptors convert the incoming photons into electrical signals and subsequent processing is begun by the interconnections between the various types of retinal cell.

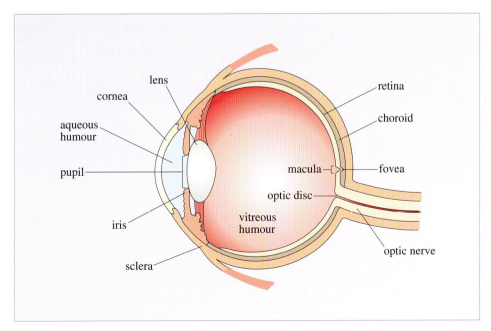

Figure 3.1   Cross section of the eye.

3.2 The pupil and the iris

When you gaze into someone's eyes you notice the colour – blue, brown, grey-green, for example. You also, although often subconsciously, notice the size of their pupils. The coloured part of the eye is the iris, a muscular diaphragm that controls the size of the pupil. Although admiring the colour of the iris is the stuff of many a romantic encounter, from the point of view of the optics of the eye it is the size of the pupil that is important! In the normal eye the diameter of the pupil can vary between 2 and 8 mm.

The pupil has a reflex response to illumination – it constricts rapidly if the luminance is increased and dilates when the luminance is decreased. If the light falling on one eye is increased then the pupil of that eye constricts. This is known as the **direct light reflex**. At the same time the pupil in the other eye also constricts; this is the **consensual light reflex**. If you can find a willing victim you might like to try this out: with your subject having been in a darkened room for several minutes, shine a torch into one eye (make sure it is not too bright) and watch how the pupils of both eyes contract.

Having a pupil of varying size has three main advantages:

- It modifies the amount of light entering the eye.
- The **depth of field** is improved when the level of illumination is high.
- The effect of any **aberrations** (defects) in the lens is also reduced under high luminance conditions.

All these points will be discussed again in Section 3.7 where we consider the effects of changing the level of illumination (see also Box 3.1).

The pupil also decreases in diameter when the eyes converge to look at a close object. This is referred to as the **near reflex**.

Box 3.1 An interesting aside: the emotional effect of pupil size

The pupil is under the control of the autonomic nervous system, so not amenable to voluntary control. Instead it can be affected by emotional state (over and above the principal impact of light). It has been shown that photographs of faces with artificially enlarged pupils are judged to be more attractive than the originals. The judges are typically unaware of what is affecting their judgement. A person's pupils tend to dilate when they find another person attractive, so presumably we are able, albeit subconsciously, to recognize and respond to this.

According to legend, the old Chinese jade merchants knew that during negotiation, a customer's pupils would change size when the price they were prepared to pay was reached. The merchant would then refuse to budge, no matter how much the customer tried to haggle, knowing that he would eventually win.

3.3 The dioptric apparatus

3.3.1 The eye as a camera

One can draw a simple parallel between the dioptric apparatus of the eye and a camera. The structures of both consist of a light-tight container, an adjustable aperture and a converging lens that focuses the image onto the detector. Figure 3.2 compares the components of the eye and of a simple camera. However, as we shall see, this is a rather simplistic picture of the eye. For a start, the lens in most cameras has a fixed focal length (see Box 3.2) and the focusing is achieved by moving the lens so that the lens–film distance is altered, whereas the lens in the eye is adjustable (the focal length can be varied) but the lens–retina distance is fixed.

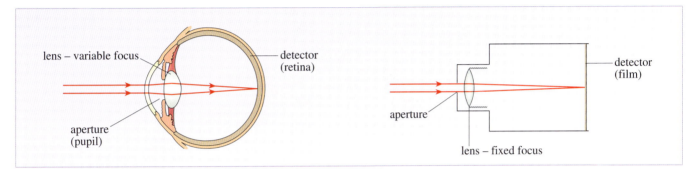

Figure 3.2   A comparison between the eye and a simple camera. Both have a variable aperture, a focusing lens and a detector system. In both cases there is also a 'shutter' (not shown) which can exclude light altogether – in the case of the eye this is the eyelid.

3.3.2 The focusing power of the eye

Looking at Figure 3.2 you would think that it is the lens of the eye that focuses the incoming light, and indeed this is the popularly held view. Closer examination of the refracting surfaces and the refractive indices of the materials involved reveals that this is not the whole story. To look at this more precisely we need to start with some of the basic physics of **refraction**. Boxes 3.2 and 3.3 (overleaf) give a reminder about refraction and refractive index and about the way in which lenses can focus light. After studying these try to answer the following question.

○ The eye needs to be able to focus light from a source at infinity to a sharp image on the retina. If the average distance between the centre of the lens system and the centre of the retina is 16.7 mm, what is the required focal length of the lens system? What is the power of this lens system?

● Since a lens focuses light from a source at infinity to the focal point, the required focal length is 16.7 mm. The power of the lens (Box 3.3) is measured in dioptres (abbreviation D) where:

>   power in dioptres = 1/focal length (in metres)

Therefore the required power is $1/16.7 \times 10^{-3} \approx 60\,\mathrm{D}$.

This is a very high value. By comparison, those of us who wear glasses for short or long sight will usually have a prescription for only a few dioptres. On closer examination it turns out that the power of the eye lens itself is only about 20 D.

Box 3.2 Refraction

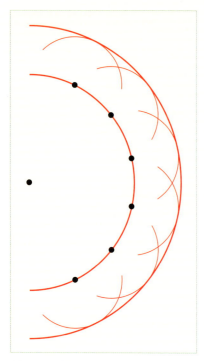

Figure 3.3 Each point on the wavefront can be considered as a new source of waves. The effect is familiar to all those who enjoy throwing stones into ponds!

We have already established that light can be treated as a wave and that the speed of that wave is reduced when it enters any medium other than air. The science of optics concerns itself largely with the behaviour of these light waves at the boundaries between different media. One very useful way to examine this behaviour is to use **wavefront** constructions, an idea first developed by the seventeenth-century Dutch scientist Christiaan Huygens. His suggestion was that each point on a wavefront could be treated as a secondary source of spherical waves, as shown in Figure 3.3.

If we consider a plane wavefront and apply **Huygens' principle** then the net effect is shown in Figure 3.4. The **principle of superposition** can now be applied – the amplitudes of all the waves at a particular point at a particular time are added together. The result of this is a new wavefront. In the case of a plane wave in a continuous medium, illustrated here, the wave has simply moved forward – exactly as we would have expected from everyday experience! The speed of that progression is the speed of the wave in that medium, the distance travelled in time t being vt where v is the speed.

However the behaviour of the wavefront is different if it crosses into another medium where the speed is different. Look at the wavefront labelled ABC in Figure 3.5a and think of this as a source of secondary spherical waves.

As the speed in medium 2 is lower than in medium 1, the spherical waves from A will not travel as far in time t as those from C which are still in medium 1. So after time t the new wavefront looks like A'B'C'. The effect of entering the new medium has been to turn the wavefront slightly.

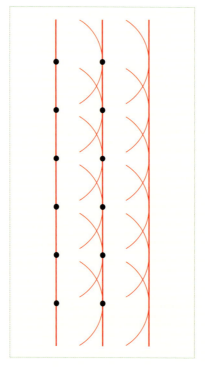

Figure 3.4 Huygens' principle applied to a plane wave.

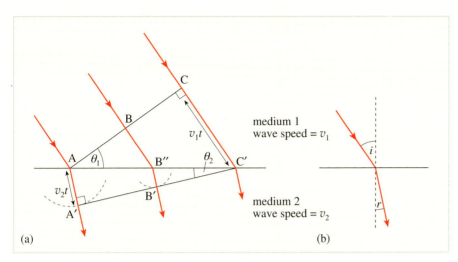

Figure 3.5 (a) The Huygens' approach to refraction. The speed in medium 2 (v_2) is less than the speed in medium 1 (v_1). This causes the wavefront to be refracted. (b) The ray representation of refraction gives a simpler picture. Note that $i = \theta_1$ and $r = \theta_2$.

The amount of turning depends on the speeds in the two media and it is not difficult to show geometrically that:

$$\frac{\sin \theta_1}{\sin \theta_2} = \frac{v_1}{v_2} \tag{3.1}$$

Wavefronts are a useful concept for trying to understand refraction but they can become a little cumbersome when drawing diagrams. It is usually easier to draw a 'ray' of light such as one of the lines AA′, BB″ or CC′ that is normal (lies at right angles) to the wavefront. When calculating the amount of refraction that occurs, the angles used are the angles between the ray and the normal. These are shown in Figure 3.5b, and as the angle between the wavefront and the surface is the same as the angle between the normal to the wavefront (the ray) and the normal to the surface, $i = \theta_1$ and $r = \theta_2$, so we can write:

$$\frac{\sin i}{\sin r} = \frac{\sin \theta_1}{\sin \theta_2} = \frac{v_1}{v_2} \tag{3.2}$$

This is known as **Snell's law of refraction**.

As the speed of light in a vacuum is an internationally defined constant, given the symbol c, the usual practice is not to quote the actual speed in other media but to express that speed in terms of c. The **refractive index** of a material, n_{medium}, is defined by

$$n_{\text{medium}} = c/v_{\text{medium}} \tag{3.3}$$

So Snell's law may now be written:

$$\frac{\sin i}{\sin r} = \frac{n_2}{n_1} \tag{3.4}$$

Refractive index is different for different materials (for air, as the velocity of light is almost equal to that in a vacuum, $n_{\text{air}} = 1$) and also for different wavelengths – a point we shall return to later.

Box 3.3 Lenses

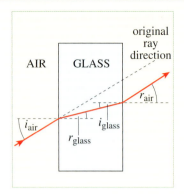

Figure 3.6 The passage of a ray of light through a parallel-sided block results in a lateral displacement but no net change in direction.

In Box 3.2 we saw that a light ray is refracted when it enters a different medium. In this section we shall consider what happens when a light ray passes through several different surfaces in succession.

The passage of a ray of light through a parallel-sided block of glass (refractive index n_{glass}) is shown in Figure 3.6.

At the first interface the ray is refracted towards the normal and the angle of refraction, r_{glass}, is determined by Snell's law. Using the notation of the diagram we get:

$$\frac{\sin i_{air}}{\sin r_{glass}} = \frac{n_{glass}}{n_{air}} \qquad (3.5)$$

At the second interface the refraction is away from the normal as the second medium is air. In this case Snell's law becomes:

$$\frac{\sin i_{glass}}{\sin r_{air}} = \frac{n_{air}}{n_{glass}} \qquad (3.6)$$

Since the two sides of the block are parallel $r_{glass} = i_{glass}$ and so the ray of light leaves the glass block at the same angle as it entered. But it has been displaced laterally. This should also be clear from the diagram. A parallel-sided block of glass, therefore, has no effect on the image formed on the other side – something that we all know from looking through glass windowpanes!

Now consider a triangular prism as shown in Figure 3.7. It is clear from these diagrams that, when the two sides are not parallel, the refraction at the output will no longer be the opposite of the refraction at the input and there is an overall change of direction for the light ray. Figure 3.7 shows that the direction of the deviation depends on whether the prism has its apex above the base or vice versa. It also depends on the angle at the apex of the prism (A) and of course on the refractive index of the glass or other material used to make the prism.

Figure 3.7 The deviation of light as it passes through a prism depends on the orientation of the prism.

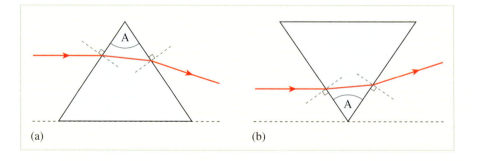

(a)   (b)

In optics it is often necessary to consider the behaviour of light passing through a lens formed by two curved surfaces. We can envisage what happens when light passes through a lens by treating it as a stack of prism-like pieces as shown in Figure 3.8. Figure 3.8a shows the case where the prisms are positioned so that all the incident rays parallel to the axis converge to a point. This is a **converging system**.

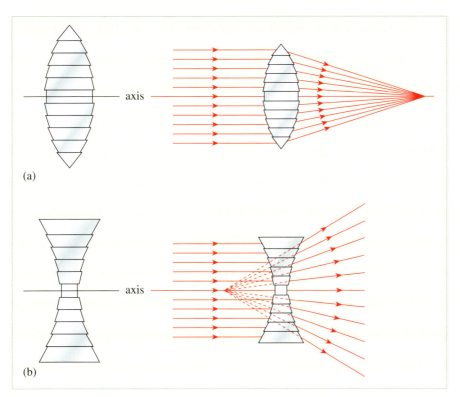

Figure 3.8 A good idea of the behaviour of a lens can be obtained by treating it as a stack of prisms. (a) Prisms stacked to give a configuration that behaves like a converging lens; (b) a configuration that behaves like a diverging lens.

The alternative arrangement is shown in Figure 3.8b. This is a **diverging system**, and all the incident parallel rays are refracted so that they appear to have diverged from one point. Now imagine that the number of individual prisms in the stacks is increased (and their size decreased proportionately) so that the change in prism angle between prisms is smaller and smaller. Eventually the surfaces become smooth curves and we could now describe the objects as lenses. In studying the eye we are going to be concerned mostly with converging systems, so we will concentrate on those.

Because this book is printed on two-dimensional paper we have so far avoided one important difference between the stack of prisms and a lens such as is found in the eye. The stack of prisms is strictly speaking a cylindrical lens in that it only refracts the light in one plane – it has no horizontal focusing effect. This is illustrated in Figure 3.9 (overleaf).

The eye lens has approximately the same cross-sectional profile in any plane perpendicular to the axis (as do most other lenses). Such a lens will bring all incident light parallel to the axis to a focus at one point. This point is known as the **focal point** and the distance between the centre of the lens system and the focal point is known as the **focal length**. (Incidentally any parallel rays that are at an angle to the axis will be brought to a focus at a point in the plane that is perpendicular to the axis and passes through the focal point – this is known as the **focal plane**.)

The focal length of a lens is a very important quantity. It is expressed as a positive number for a converging lens and a negative number for a diverging

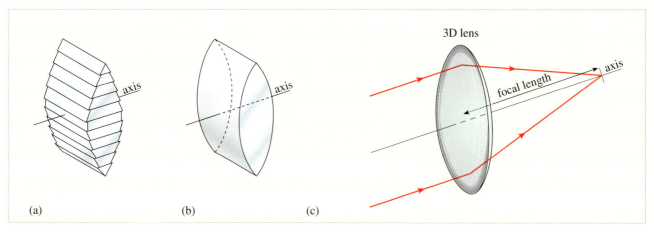

Figure 3.9 The 'stack of prisms' model (a) only behaves like a lens (b) for the light in one plane. A lens with spherical surfaces (c) refracts light in all directions. The focal length is the distance between the centre of the lens and the point at which light parallel to the axis is focused.

lens. An alternative way of expressing the ability of a lens to converge or diverge light is to quote its **power**. This is defined by:

power = 1/focal length (in metres)

and the units of power are **dioptres** (abbreviation **D**). A lens with a high positive value of power is a converging lens with a short focal length. A lens with a negative power value is a diverging lens.

So where does the rest of the focusing take place? To calculate this it is better to consider the refraction that occurs at each curved surface rather than to look at the cornea and the lens as separate entities. The lenses considered in Box 3.3 were glass lenses in air. In the case of the eye we have a more complicated system with layers of material with different refractive indices. For light entering the eye these surfaces are shown in the schematic diagram in Figure 3.10.

Figure 3.10 The refracting surfaces of the eye. Note that the lens is actually a complex layered structure but has been simplified for the purposes of illustration.

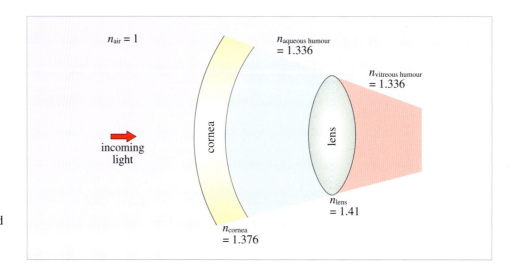

The approximate **refractive power**, P, of each surface can be calculated using the equation:

$$P = (n_2 - n_1)/r \tag{3.7}$$

where n_1 and n_2 are the refractive indices of the first and second media, and r is the radius of curvature of the surface (in metres). Note that r is negative for the posterior surface of the lens as it curves in the opposite direction from the other surfaces.

○ Use the information on Figure 3.10 and in Table 3.1 to calculate the missing power values in the table. The refractive index of air may be taken as 1. (Remember that if the second medium has a lower refractive index and the power is then negative, this indicates divergence rather than convergence.)

Table 3.1 Calculation of the power of the different surfaces of the eye.

Surface	Radius of curvature/mm	Power
anterior corneal surface	7.8	$P = (1.376 - 1)/7.8 \times 10^{-3} = 48.2\,\mathrm{D}$
posterior corneal surface	6.3	$P = (1.336 - 1.376)/6.3 \times 10^{-3} =$
anterior lens surface	10.0*	$P =$
posterior lens surface	-6.0	$P =$

* For the eye focused at infinity.

● You should obtain values of $-6.3\,\mathrm{D}$, $7.4\,\mathrm{D}$ and $12.3\,\mathrm{D}$ for the three remaining surfaces. An estimate of the total power as $62\,\mathrm{D}$ can be obtained by adding all the values. This does not give a completely correct value as the surfaces are slightly separated from each other (the correct equation is given in Chapter 9 of the Reader, *The cornea*). However the figures in this table should be enough to convince you of the important role the air/cornea surface plays in forming the image on the retina.

3.3.3 Focusing the lens

In Section 3.3.2 you calculated the power that the dioptric system needed to have in order to focus light from an infinite source onto the retina. The most distant point on which the eye can focus is known as the **far point**. But the eye can also focus on close objects. For most people the **least distance of distinct vision** or **near point** is about 250 mm from the lens. If the lens power remained at about 60 D then it might be expected that, since the object is closer to the eye, the image would be formed behind the retina, as shown in Figure 3.11.

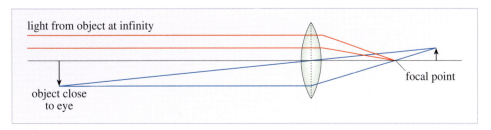

Figure 3.11 For a converging lens of fixed focal length the image of an object which is at infinity will be at the focal point; the image of an object closer than that will be further from the lens.

There is a simple equation relating the object distance, u, the image distance, v, and the focal length, f, of a lens. It is:

$$\frac{1}{u} + \frac{1}{v} = \frac{1}{f}$$ (3.8)

This equation is known as the **lens equation**. Strictly speaking it is only true for thin lenses and for rays close to the centre of the lens, but it will be sufficiently accurate to allow us to estimate image and object distances for the eye.

○ Use equation 3.8 to calculate the image distance for an object 250 mm from a lens of power 60 D.

● Rearranging the equation gives:

$$\frac{1}{v} = \frac{1}{f} - \frac{1}{u}$$

$1/f$ is 60 D, or m^{-1}, and, if u is 0.250 m then $1/u = 4$ m^{-1}. Therefore $1/v = 60 - 4 = 56$ m^{-1}. So $v = (1/56)$ m $= 17.8$ mm. If the eye does not alter the focal length of the dioptric system then the image will be approximately 1 mm behind the retina and therefore out of focus.

As most of us can see close objects quite clearly, the power of the dioptric system must change to allow focusing of close objects. Another calculation using the lens formula quoted above shows that, for an object at 250 mm to be focused on a retina 16.7 mm from the lens, the power of the dioptric system needs to be about 64 D, a significant increase in power. While it is not true to say that all the focusing in the eye takes place in the lens, it is true that, for mammals, altering the power of the *lens* is the only way in which the focus of the eye can be rapidly changed so that close objects can be seen.

The shape of the lens is changed by altering the tension in the zonular fibres. If the fibres are taut then the lens is thinner – this corresponds to the eye being focused for distant vision. If the zonular fibres are more relaxed then the lens becomes 'fatter' and more powerful and the condition for close vision can be met. The muscular mechanisms are complicated but it is important to note that the ciliary muscle controls the tension in the zonular fibres; when the ciliary muscle (Figure 3.12) is relaxed the zonular fibres are taut and the eye is focused for distant vision. Focusing on close objects requires the ciliary muscle to contract and is therefore *not* the relaxed condition for the eye.

Figure 3.12 When the ciliary muscle is relaxed, the zonular fibres are taut, causing the lens to become flat. When the ciliary muscle contracts, the zonular fibres are relaxed and the lens becomes fatter.

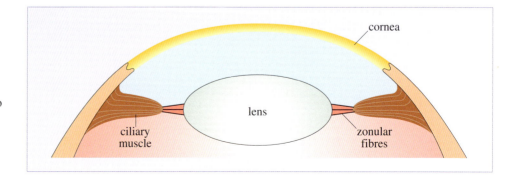

An interesting puzzle concerning this process of **accommodation** is the question of how the eye knows which way to alter the focal length. When we change from looking at a distant object to looking at a close one we are unaware of the accommodation taking place – it is clearly a reflex reaction. So what are the signals that tell the brain which way to alter the lens? The increase in convergence, or the inward rotation of the eyes in order to focus on a close object, is one possible clue that the brain may have. Other suggestions have been related to aberrations in the lens (see Section 3.7.1) but the matter remains under discussion.

Activity 3.1 Refraction and lenses

You should now go to the CD-ROM where you will find some helpful material on refraction and lenses. Further details are given in the Block 4 *Study File*.

3.3.4 Refractive defects of the eye

You may be wearing spectacles to read this course material. Or you may need to wear spectacles to see clearly in the distance. Or both! The study of refractive defects of the eye is an enormous subject and we can do no more than give outlines of two of the most common defects and their correction.

If the refractive power of the eye is too strong then the image of a distant object is formed *in front of* the retina. This can happen because there is too much refractive power, or because the front–back distance of the eye is too long. Someone with this defect will be unable to focus clearly on distant objects, although they may well have no problems with close-up objects – they are short-sighted (or have **myopia**, to use the technical term). Figure 3.13 shows a ray diagram for myopia.

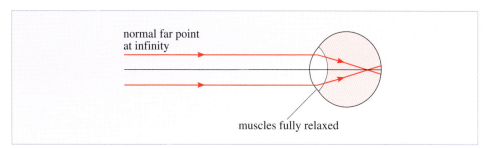

Figure 3.13 The myopic eye uncorrected.

The other simple defect that may occur is the opposite – the refractive power of the eye is, for whatever reason, insufficient to focus the image on the retina; instead the image is formed *behind* the retina. This problem is worst when looking at close objects so such people are said to be long-sighted (or have **hyperopia**). Figure 3.14 shows this effect.

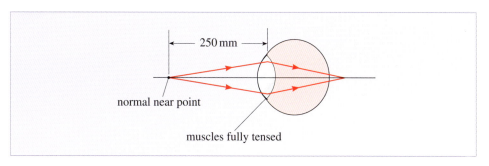

Figure 3.14 The hyperopic eye uncorrected.

It is possible that you may be reading this and thinking that you have both of these complaints! In that case you are probably over forty years of age and have **presbyopia**. As the lens ages it becomes less able to change its shape to accommodate in the manner described in Section 3.3.3. The result is that both distant objects and close objects become blurred.

So how can these defects be corrected using either spectacles or contact lenses? For myopia the correcting lens must cause the rays reaching the eye to diverge so that they focus on the retina instead of in front of it. The way this is done, using a diverging lens, is shown in Figure 3.15a. A diverging lens has a negative power, so when this is subtracted from the power of the eye, which is too large, the required correction is achieved. The hyperopic eye, on the other hand, is corrected with a converging lens as in Figure 3.15b. This has a positive power, which adds to the insufficient power of the eye to give an increased power.

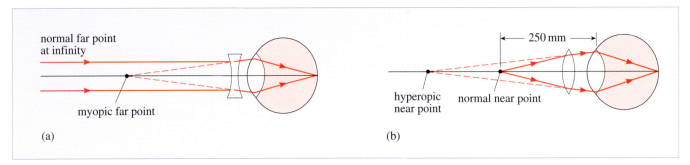

(a)

(b)

Figure 3.15   Correcting (a) the myopic eye with a diverging lens, and (b) the hyperopic eye with a converging lens.

And what if you have presbyopia? In that case you probably know the answer – bifocal (or varifocal) lenses, which act as concave lenses for distance and convex lenses close up.

3.3.5 The special properties of the cornea

From what has been said so far it should be clear that the cornea is a very important part of the eye. It is tough enough to protect the rest of the eye from injury and also carries out about two-thirds of the focusing of visible light in the eye. The latter requires the cornea to have a very smooth optical surface, a precisely defined curvature and, most of all, to be 100% transparent.

One of the key reasons for the transparency of the cornea is due to the phenomenon of diffraction. If you are not familiar with this concept you might like to read Box 3.4 (overleaf) on diffraction before reading the next chapter of the Reader. Alternatively, you may prefer to read the chapter first and then come back to the box to try to understand the basic ideas of diffraction.

You should now read Chapter 9 of the Reader, *The cornea* by Keith Meek, to find out how the special properties of the cornea are achieved.

Question 3.1

List some of the special, and perhaps surprising, properties of the cornea.

Question 3.2

Why is it not possible to see clearly underwater without swimming goggles?

Question 3.3

Why is the cornea transparent to visible light, yet a diffraction pattern is obtained when X-rays are passed through corneal tissue?

Question 3.4

One member of the Course Team wears spectacles. His prescription gives the power of the lenses as:

left eye: −1.25 D   *right eye*: −1.50 D

(a) What refractive defect does he have?

(b) When does he need to wear his spectacles?

Summary of Sections 3.1–3.3

Light entering the eye passes through a dioptric system of cornea and lens in the midst of which is a pupil of adjustable diameter. The dioptric system forms an image of the scene observed on the retina at the back of the eye.

The pupil is a circular aperture in the coloured iris. Its diameter can be varied from 2 mm to 8 mm. As well as modifying the amount of light entering the eye, it also alters the depth of field and the effect of aberrations.

The focusing of light in the eye can be described using the usual rules of geometrical optics. Refraction takes places at curved interfaces between media of different refractive indices. The approximate power of each surface is given by the equation:

$P = (n_2 − n_1)/r$

The total power of the relaxed eye is of the order of 60 D.

The cornea has special properties that allow it to be tough, deformable and almost totally transparent. The latter depends on the fact that the spacing of the collagen fibrils in the cornea is less than the wavelength of light.

For the normal eye the least distance of distinct vision, or near point, is about 250 mm. The (approximately true) thin lens equation:

$$\frac{1}{u} + \frac{1}{v} = \frac{1}{f}$$

can be used to show that this corresponds to a power of 64 D. The change in power, or accommodation, is achieved by alterations in the ciliary muscles and the zonular fibres. The relaxed condition for the eye is when it is focused at infinity.

Refractive defects of the eye such as myopia and hyperopia can be corrected using concave or convex lenses respectively.

Box 3.4 Diffraction

Huygens' principle was used in Section 3.3.2 to examine the behaviour of light waves at an interface between two different materials. Huygens' principle can also be used to describe the phenomenon of **diffraction**.

Consider first a pair of narrow slits illuminated by a plane wavefront. Huygens' principle tells us that each point on the wavefront can be treated as a secondary source of spherical waves. Figure 3.16 shows this principle applied to the narrow slits. This diagram clearly shows that the light does not just propagate in

the forward direction – it also spreads out sideways. This spreading is known as diffraction.

The principle of superposition can now be used to find the amplitude of the combined waves at points some distance from the slits. At any angle the total amplitude can be found by adding the amplitudes of all the waves travelling in that direction, but taking into account their phase. Figure 3.17 shows three examples. In this case two waves of the same amplitude and frequency are added. In Figure 3.17a the waves are in phase so the resultant amplitude is large. This is known as

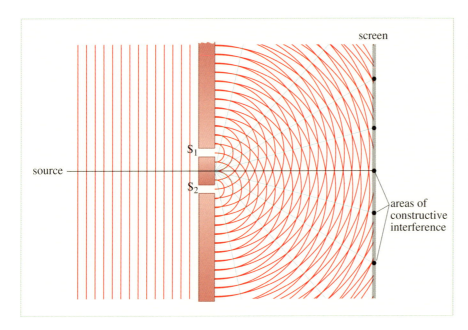

Figure 3.16 Using Huygens' principle each slit is treated as a secondary source of spherical waves. The principle of superposition can be used to find the amplitude of the resultant wave at different angles.

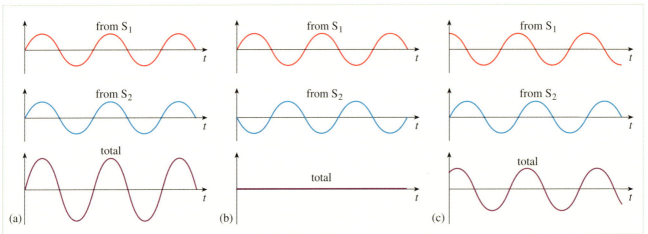

Figure 3.17 The resultant amplitude is found by adding the amplitude of all the different wavefronts reaching a point. (a) The waves are in phase; (b) they are half a cycle out of phase; (c) an intermediate situation.

constructive interference. In Figure 3.17b the waves are half a cycle out of phase so the resultant amplitude is zero. This is known as **destructive interference**. Figure 3.17c represents an intermediate case where the resultant wave has an amplitude slightly larger than the individual original waves.

Figure 3.18 can be used to look more closely at the phase difference between the waves from adjacent slits. As all the wavefronts are in phase when they leave the slit, the relative phase of any two waves at angle θ depends on the different distances travelled by the wavefronts.

The wavefronts that have started from slits S_1 and S_2 have travelled different distances. The difference in their distance can be worked out from simple geometry:

$$\text{path difference} = d \sin \theta \qquad (3.9)$$

If the angle $\theta = 0$, which is the case straight ahead of the slit, all the wavefronts have travelled the same distance and therefore they will all be in phase, giving

a very high intensity. At any other angle where the path difference is a multiple of the wavelength, λ, then the wavefronts from S_1 and S_2 will also add up to give a high intensity. If the path difference is an odd multiple of $\lambda/2$ then the wavefronts will cancel out, giving a minimum amplitude. So the condition for a maximum amplitude is:

$$n\lambda = d \sin \theta \quad \text{where } n = 0, 1, 2, \dots\dots \qquad (3.10)$$

Diffraction will occur whenever the wavelength, λ, is comparable to the slit spacing d. If $\lambda < d$ then there will be a bright spot at $\theta = 0$ (known as the zeroth order as $n = 0$). The next bright region, where $n = 1$, will be at the angle given by:

$$\sin \theta = \lambda/d$$

However, if $\lambda > d$ then there will only be the zeroth order bright spot as the value of $\sin \theta$ cannot exceed 1. As you will see from Chapter 9 of the Reader, this is an important factor determining the properties of the cornea.

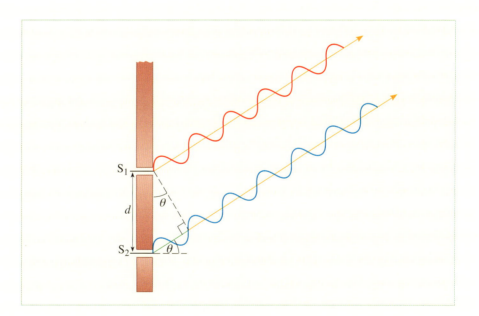

Figure 3.18   The geometry of diffraction. When the path difference between the wavefronts from S_1 and S_2 (green line) is equal to a whole number of wavelengths there will be a bright region.

Activity 3.2 Diffraction

If you are not familiar with the idea of diffraction you should now undertake the activities provided on the CD-ROM. Further details are given in the Block 4 *Study File*.

STUDY
FILE

3.4 The retina

We have been discussing the formation of an image on the retina and it is now time to investigate the structure of the retina. The retina is actually a part of the brain and its complex structure has evolved to serve two different types of function. First, it provides a receptive surface where light energy is transduced (converted) into a neural response. This is achieved by specialized photoreceptors. Second, it begins processing and organizing the initial neural response so that appropriate information is relayed in the most useful form to higher visual centres. This is achieved by a complex network of interconnections between a wide variety of specialized cells.

3.4.1 Retinal transduction

Figure 3.19 is a photomicrograph of a cross section of human retina, and Figure 3.20 is a simplified schematic diagram, which makes the structure clearer. Familiarize yourself with the overall structure and locate the **photoreceptors**, which are the first proper stage of neural processing. There are two types of photoreceptor – **rods** and **cones** – which have rather different roles. Rods are very sensitive and function best under dim (**scotopic**) conditions; they are well adapted for night vision. Cones are less sensitive and function best under bright (**photopic**) conditions. In effect they form a separate retinal system for operating in daylight. Rods outnumber cones by 20:1 and are 1000 times more sensitive to light. For the moment we can consider all photoreceptors as essentially the same; they are specialized structures for transducing light energy into the electrochemical energy that is the first stage of the neural response. Even though, as Figure 3.20 shows, light passes through several interneurons before reaching the receptors, these interneurons are not specialized for transduction and so do not respond to light. They only respond indirectly as a result of the neural activity that the light produces in the receptors.

Each receptor measures the amount of light at one point in the image, so the pattern of neural activity over many receptors forms a kind of **neural image**, which directly mirrors the pattern of light in the image, only in units of neural response rather than units of light.

Retinal receptors, as illustrated for a rod in Figure 3.21, have an outer segment made up of a pile of stacked membranes, and an inner segment containing the nucleus and synaptic contacts. The membrane stacks contain photosensitive pigment which is

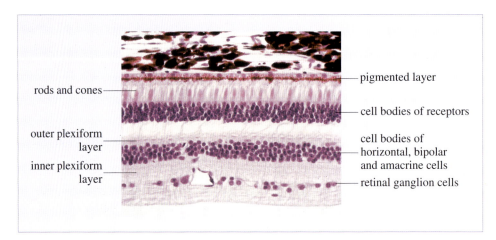

Figure 3.19   Photomicrograph of a cross section of vertebrate retina.

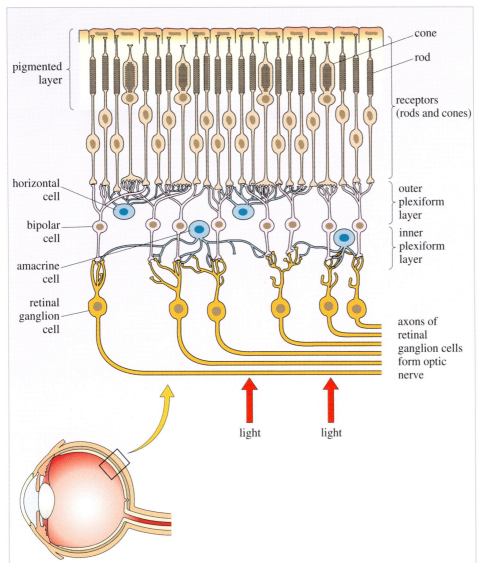

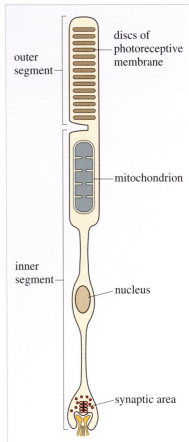

Figure 3.21   Diagram of a rod photoreceptor.

Figure 3.20   Schematic cross section of vertebrate retina.

broken down, or bleached, by light. This breakdown changes the cell's permeability to ions, thus producing a receptor potential (Block 2). Once bleached, the pigment is regenerated so that the process can be repeated (Box 3.5 overleaf).

The process of depolarization and hyperpolarization described in Box 3.5 is exquisitely sensitive – a very small amount of light produces a measurable receptor potential.

You should recall from Section 3.3 of Block 2 that the depolarization signal of an action potential generally causes an increase in transmitter release, while hyperpolarization causes a decrease. Consequently, retinal receptors actually release more transmitter in the dark than in the light. This may seem strange – but remember that a darkening in the field of view (e.g. a shadow) is often a more relevant stimulus than a lightening.

Box 3.5　The photochemistry of vision

The initial event in vision is the absorption of a photon of light by the visual pigment present in rods and cones. The pigment in rods is called **rhodopsin**, each molecule of which consists of a protein **opsin** to which is attached a molecule called **11-*cis*-retinal** (Figure 3.22a). The pigments in the three types of cone, generically called **iodopsin**, also contain 11-*cis*-retinal, but it is attached to three different, though related, opsin molecules. The energy of the absorbed photon converts 11-*cis*-retinal into its so-called isomer **all-*trans*-retinal** (Figure 3.22b).

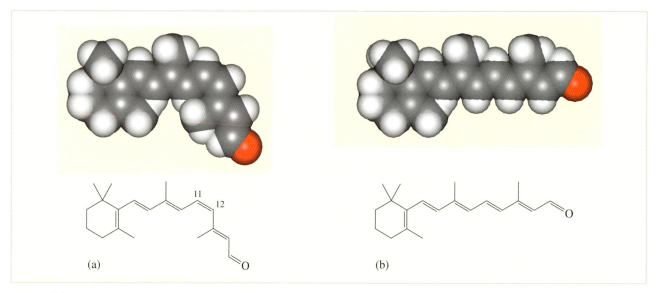

(a)　　　　　　　　　　　　　　(b)

Figure 3.22　(a) 11-*cis*-retinal as a space-filling model (upper) and a skeletal formula (lower); (b) all-*trans*-retinal as a space-filling model (upper) and a skeletal formula (lower).

Most large molecules are 'floppy' and can change their shape millions of times a second at room temperature. This happens because rotation of one part of the molecule relative to the other around each single bond occurs very easily. In contrast, the presence of a double bond makes that part of the molecule rigid. Consequently the two structures corresponding to rotation about the double bond represent different substances, since they do not interconvert without the application of a substantial amount of energy, e.g. with very high temperatures or by the use of light.

One of the simplest examples of this phenomenon is shown in Figure 3.23. The two molecules shown are called **isomers** since, though they are different substances, they have the same molecular formula. They are made from different arrangements of exactly the same set of atoms. This similarity is indicated in the names of the substances. They are called *cis* or *trans* depending whether the two parts of the molecule attached at either end of the double bond are on the same side (*cis*) or diagonally opposite (*trans*). Hence the names 11-*cis*-retinal and all-*trans*-retinal (Figure 3.22). The number 11 in the *cis* isomer denotes the position of the crucial double bond; in the other isomer, all the double bonds in the side

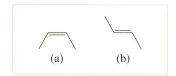

(a)　　　　　(b)

Figure 3.23　(a) *cis*-but-2-ene; (b) *trans*-but-2-ene.

chain are *trans*. The geometry of the other double bonds present in the molecule is not affected by light absorption. (This pattern of alternating single and double bonds is often called a **polyene chain**.)

The absorption properties of rhodopsin and the three cone pigments in the visible region of the spectrum are largely due not to the opsin portion of the molecules, which is colourless in all four cases, but to the retinal portion. It is the presence of the alternate pattern of single and double bonds that gives rise to the visible light absorption. In this context the retinal portion is said to be the **chromophore**. The four different opsins do, however, have the effect of modifying the wavelengths that the four pigments absorb, and hence giving rise to their different colours.

The opsins are membrane proteins, with the middle of the molecule embedded in the lipid bilayer. The protein chain traverses the lipid bilayer seven times, such that the hydrophilic ('water-loving') amino acid residues predominate at the membrane surfaces and the hydrophobic ('water-hating') residues are predominantly in the lipid bilayer. The 11-*cis*-retinal is attached to the side chain of lysine 296, located midway in the membrane-spanning seventh helix (Figure 3.24).

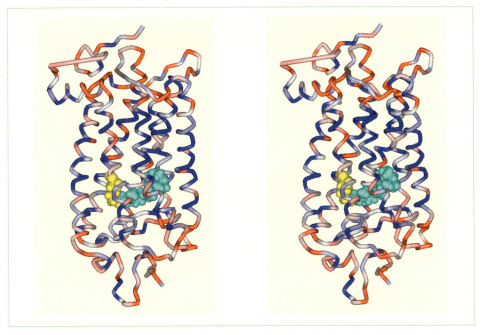

Figure 3.24   Stereoscopic model of rhodopsin showing the location of the bound 11-*cis*-retinal molecule. (You should use the stereoscopic viewer for this figure.) The retinal molecule (cyan) and the amino acid in the opsin to which it is attached (lysine; yellow) are shown as space-filling models; just the backbone of the rest of the opsin molecule is shown for clarity. The rhodopsin has three domains or regions: extracellular (lower), transmembrane (middle) and cytoplasmic (interior). Note the predominance of hydrophobic amino acids (blue) in the transmembrane domain of the opsin chain with a greater proportion of polar amino acids in the extracellular and cytoplasmic domains.

As mentioned above, *cis* and *trans* isomers can interconvert if enough energy is applied, and it is the conversion of 11-*cis*-retinal to all-*trans*-retinal brought about by light absorption that is the key step in visual transduction. The conversion of 11-*cis*-retinal to all-*trans*-retinal causes a change in the shape of the opsin molecule and triggers a cascade of events, which includes bleaching of the rhodopsin or iodopsin, activation of a G-protein called transducin, and release of cyclic-GMP, which closes cGMP-gated cation channels in the photoreceptor membrane. In the dark, a steady current flows into the open channels, carried mainly by Na⁺ ions, constituting a **dark current** that partially depolarizes the photoreceptor cell. On light stimulation, closure of these channels stops the dark current, causing the photoreceptor cell membrane to hyperpolarize and cease neurotransmitter release to the associated neurons. The free all-*trans*-retinal released in the bleaching process is in due course converted back into the *cis* form by a series of enzyme-catalysed reactions, whereupon it reattaches to another opsin molecule, ready to begin the process again.

As with all such cycles, efficiency is less than 100% and the retinal needs constant replenishment. 11-*cis*-retinal is made from vitamin A. This accounts for our need for this vitamin, without which you would eventually lose your sight, though it has other biological functions as well. Vitamin A occurs in a number of animal products (for example, fish liver oils); otherwise, it is made from certain members of a family of compounds called **carotenes** (Figure 3.25). The best known of these is *beta*-carotene, which is found in carrots and green vegetables, for example. So keep eating your greens!

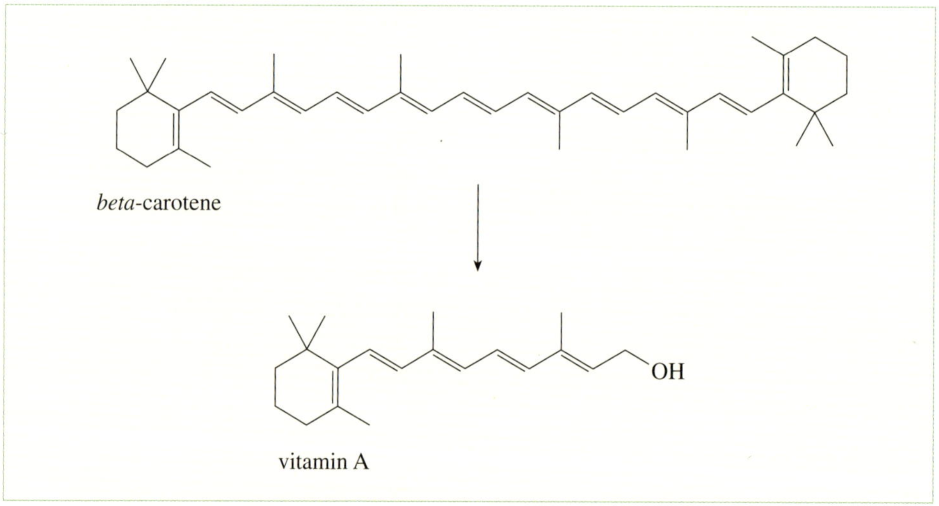

beta-carotene

vitamin A

Figure 3.25   Formation of vitamin A from *beta*-carotene.

At this point you should read up to the end of Section 3 of Chapter 10 of the Reader, *The retina* by Jim Bowmaker. Some of the chemistry is quite challenging, so if you do not have much prior knowledge of organic chemistry you should aim to understand the basic ideas and not spend too much time on the detail.

3.4.2 Receptive fields

So far, the visual process seems relatively straightforward. Transduction converts a pattern of light into a corresponding pattern of neural response and all the spatial relationships in the original image are preserved in the resulting neural image. Rather like a photograph in a newspaper, the neural image is made up of small points, each point being the response of a single receptor. Since the receptors are very small and closely packed, there are many small points and the neural image provides a faithful representation of the original pattern of light.

○ The point-like nature of the image as represented by the responses of retinal receptors is likened to a photograph in a newspaper. In what important respect does the neural response differ from the photograph?

● Retinal receptors respond more to dark regions of the image and less to light regions in terms of the amount of transmitter released, so their pattern of response is more like a photograph negative than the photograph itself.

Although the notion of a neural image is useful, it can be misleading if taken too far. A faithful neural representation of the image is a good place to *start* visual processing, but the purpose of visual processing is not just to take some sort of neural photograph. Remember that vision is not only about describing the image, but also about working out how it might have been produced. There are very few situations where we scan a static scene. Mostly we use vision to negotiate and interact with the physical environment. The eye has evolved in this context rather than as an aid to enjoyment of a picture in an art gallery! The function of the early stages in visual processing is thus to extract information that is particularly relevant to that interpretative task, rather than simply to preserve the original image. This point is brought home by the very next steps in the visual process, which also take place in the retina.

The **receptive field** of a visual cell is the region of the image to which the cell responds. As implied above, the receptive fields of retinal photoreceptors are very small, so that the photoreceptors can provide a faithful representation of the image. However, as is clear from Figure 3.20, higher visual centres do not have direct access to the responses of individual retinal photoreceptors. Instead, the photoreceptors are connected by a complex network of interneurons to the **retinal ganglion cells** and it is these cells that project to the visual cortex. Higher visual centres thus 'see' only the output of the retinal ganglion cells. When we study the receptive fields of retinal ganglion cells we find that they have two properties that, at first, seem puzzling.

1 The receptive fields are generally much larger than those of retinal photoreceptors, so the output from the retina seems relatively coarse, lacking the fine detail provided by the photoreceptors.

2 The receptive fields are divided into separate, antagonistic sub-regions so that one region of the image excites the retinal ganglion cell whereas the neighbouring region inhibits it. Consequently, when the whole receptive field is evenly illuminated, the retinal ganglion cell does not respond at all.

Why should the retina be wired up so that it seems to throw away spatial detail and to be unresponsive to uniform illumination?

The answer to the first question, about losing spatial detail, is to do with increasing the sensitivity of the retina in dim lighting conditions. You already know that the retina contains rod photoreceptors that are specialized for dim conditions. Chapter 10 of the Reader, *The retina*, indicated that a single photon is enough to produce a measurable effect in a rod and that this response is amplified by a cascade of chemical processes. But even this is not enough to account for our exquisite sensitivity to light. Rods also summate the effects of all photons received over a 100 ms time period. This means they respond more slowly than cones and will not detect that a light is flickering if the flicker is above 12 Hz. Consequently, your temporal resolution is less good in dim light than in good light (when cones are active) which is the reason for 'light stopped play' in fast ball games. More important for the current argument, sensitivity is also increased by combining the outputs of a number of rods. Many rods typically converge onto a single bipolar cell, which in turn influences a retinal ganglion cell through two types of intermediary amacrine cells, as will be described later in Chapter 10. Because the retinal ganglion cell receives input from many rods spread over a relatively large region of the retina, it sacrifices the ability to resolve fine spatial detail. But the benefit of this arrangement is that retinal ganglion cells' responses are more sensitive and more reliable than the responses of individual photoreceptors.

The answer to the second question, about lack of response to uniform illumination, is a little more complex and we shall return to this topic in Section 4. But briefly, the retina does not need to respond to uniformly-illuminated regions of the image because they convey little useful information. Rather, the retina only needs to signal the places in the image where the amount of light changes abruptly from one point to the next, because these places correspond to the edges of objects in the world. In general, informative change is much more important than uninformative uniformity. And retinal ganglion cells are wired up to emphasize change and to ignore uniformity. How is this done?

Retinal ganglion cells have circular receptive fields. When light falls on the central portion of the receptive field the retinal ganglion cell (rgc) responds in a way that is exactly the opposite of the way that it responds when light falls on the annular periphery. The rgc is described as being either an **ON-centre** type or an **OFF-centre** type. The ON-centre type is excited (i.e. generates action potentials more rapidly) when stimulated by light falling anywhere within the centre of its receptive field. It is inhibited (i.e. generates action potentials less frequently than its background rate of firing) when stimulated by light falling anywhere in the outer ring of its receptive field. Notice that both the excitatory and inhibitory effects on the cell's firing rate occur as the result of light stimulation. The response by the photoreceptors is the same in both cases and it is the intervening neural wiring that gives rise to the ON-centre or OFF-centre properties of the rgc.

Activity 3.3 Retinal ganglion cells

You should now go to *The Senses* CD-ROM and explore the properties of retinal ganglion cells. Further details are given in the Block 4 *Study File*.

As you may have noticed already, the retina has to meet conflicting demands. In order to preserve spatial detail, cells must 'look at' only a small region of the image. But, in order to provide sensitive and reliable signals and to emphasize informative change, cells need to combine responses from different regions of the image. The retina resolves this conflict in a variety of ingenious ways. For example, it has separate systems, based on rods and cones, for dealing with dim and bright conditions and the receptive fields that deal with dim conditions are larger than those that deal with bright conditions. Under these dim conditions, it is better to detect a signal even if you do not know precisely what it is. Similarly, in any given lighting conditions, not all receptive fields are the same size. In the foveal region, the receptive fields of retinal ganglion cells are tiny, so that not much spatial detail is lost. But, as we move from the fovea into the retinal periphery, receptive field size steadily increases. Thus the retina combines a small central region able to resolve detail with a much larger peripheral region which is more sensitive but less capable of signalling detail.

Now that you have some understanding of how retinal ganglion cells respond to light falling on the retina, and why complex interactions between retinal cells are needed, read the rest of Chapter 10 of the Reader, *The retina* by Jim Bowmaker to find out more about the detailed interconnections.

This completes your preliminary study of visual processing in the retina. You will probably be aware that we haven't told you the full story, but as yet no-one has fully unravelled the secrets of retinal processing. For example, why are there over twenty morphologically different types of amacrine cell (using at least eight different neurotransmitters)? As the answers to these, and other questions, emerge in the future you should be in a good position to understand their significance.

3.5 Colour vision

You may recall that in Section 2 of Chapter 8 of the Reader, *Colour science*, the author describes the three primary colours of light – red, green and blue – and mentions that the reason that humans can work with three primary colours is probably based on the existence of three types of cone. In Chapter 11, *Functional colour vision*, also by Stephen Westland, this idea is pursued further and the concept of colour opponency is introduced. You should read this chapter now. While doing so you may also want to refer back to Section 2.2 on reflectance. After reading the chapter try answering the following questions.

Question 3.5

Explain why the human eye cannot distinguish between monochromatic yellow light of wavelength 560 nm and a mixture of equal proportions of monochromatic lights of wavelength 545 nm and 565 nm.

Question 3.6

How do you explain the 'afterimages' effect demonstrated by Figure 1.5 in this block?

Activity 3.4 Colour vision

You should now undertake these activities on the CD-ROM, which will help your understanding of colour vision. Further details are given in the Block 4 *Study File*.

Question 3.7

Figure 3.20 shows that light passes through several interneurons before it reaches the retinal receptors. However, in Chapter 10 of the Reader, *The retina*, there is a figure that shows that this is not true for all retinal areas.

(a) Which is the figure?

(b) Describe why there are no interneurons in this area.

(c) Explain why this area is different from the rest of the retina (i.e. what is the functional relevance of the arrangement).

Question 3.8

How would our experience of colour vision be different if the eye were sensitive to the frequency of light signals in the same way that the ear is sensitive to the frequency of sound signals?

Question 3.9

The human eye has cones that are sensitive to three different wavelengths. What is it that distinguishes one type of cone from another?

Summary of Sections 3.4 and 3.5

The retina contains two types of receptor, rods and cones. There are a larger number of rods and they mainly operate under scotopic conditions, giving monochromatic vision. The cones operate under photopic conditions and are responsible for colour vision.

Light falling on these receptors bleaches the pigment (rhodopsin in rods, iodopsin in cones) leading to closure of Na^+ channels and hyperpolarization.

The responses of the photoreceptors are carried by a complex network of retinal interneurons to the retinal ganglion cells, which form the output stage of the retina. Some interneurons are inhibitory and others excitatory. As a result, retinal ganglion cells have receptive fields consisting of discrete 'ON' and 'OFF' sub-regions. This arrangement means that retinal ganglion cells respond to luminance edges in the image, rather than to regions of uniform illumination.

The ganglion cell receptive fields vary in size across the retina. They are small in the fovea, giving good spatial resolution, and larger in the retinal periphery, giving good sensitivity.

The colour properties of the visual system result from there being three types of cone in the retina, with maximum sensitivities at 420, 530 and 560 nm (about 440 nm, 545 nm and 565 nm when estimated *in vivo*). This also explains why all colour stimuli can be matched by the additive mixture of three primaries.

The ratio of the responses of each type of cone is different for each monochromatic wavelength. This allows colour discrimination. Wavelength discrimination is remarkably good – being at best around 1 nm.

The spectral power distributions of most light sources are smooth functions of wavelength and can be described by the linear sum of three functions of wavelength.

It is now established that beyond the retina the cone signals are transformed into a luminance signal and two chromatic opponent signals, red–green and yellow–blue.

There are several different types of colour deficiencies. Monochromats only have one type of cone and see everything monochromatically. There are two forms of dichromat: protanopes, who lack L-cone pigment, and deuteranopes, who lack M-cone pigment. In another type of colour deficiency, all three types of cones are present, but the peak sensitivity of either the red or green cone is shifted. In protanomalous observers the red peak is shifted towards shorter wavelengths, in deuteranomalous observers the green peak is shifted towards longer wavelengths. It is also thought that there may be some females who are tetrachromats, possessing four classes of cone pigment.

3.6 Movement of the eye

3.6.1 Range of movement

Having looked at the path of the light from the pupil to the retina, we now turn to some more general topics concerning the eye and the way it works. Movement of the eyes within the head is something that we rarely think about and yet it happens all the time, and is extremely important for maintaining the image on the fovea, for focusing on objects at different distances and for binocular vision. This movement is controlled by a series of muscles outside the orbit of the eye, which are illustrated schematically in Figure 3.26. The details of the muscles need not concern us but the possible directions of motion are important.

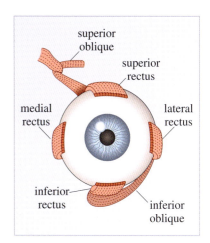

Figure 3.27 shows three possible axes of rotation. In all cases the rotation is about a point some 14 mm behind the front surface of the cornea. Rotation about the vertical axis (zz') turns the eyes from side to side. Rotation about the transverse horizontal axis (xx') elevates or depresses the eye. Rotation about the anterior/posterior axis (yy') would give rise to rolling. This last action would cause the horizontal meridian, essentially the 'horizon' seen by the eye, to rotate on the retina and upset the individual's estimate of position in space (unless there is psychological compensation). It turns out that the human eye does not generally rotate about this axis, although it rotates freely about the other two axes.

Figure 3.26 The pattern of muscles controlling the movements of the human eye. The left eye is illustrated here. The medial rectus moves the eye towards the nose; the lateral rectus moves the eye away from the nose. The superior rectus moves the eye up and the inferior rectus moves it down. The superior oblique rotates the eye so that the top of the eye moves towards the nose; the inferior oblique rotates the eye so that the top moves away from the nose.

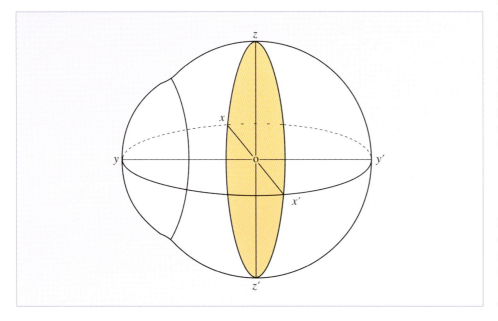

Figure 3.27 There are three axes of rotation used to define eye movements. The centre of rotation (marked O) is approximately 14 mm behind the anterior surface of the cornea.

Movement of the eyes is essential for following the motion of a moving object and for continued observation of a stationary object while the head is moving. It also plays a crucial role in binocular vision. It is always true that *the movements of the two eyes are equal and symmetric*, but there are two kinds of motion that can be distinguished. These are defined in terms of whether or not the lines between the centre of rotation and the point of fixation (known as the **fixation axes**) are the same for the two eyes. In **conjugate** (or **version**) **movements** (Figure 3.28a) the fixation axes are parallel and the movements of the two eyes are equal in all respects. In **disjunctive** (or **vergence**) **movements** (Figure 3.28b) the movements of the two eyes are equal but opposite.

Figure 3.28 (a) Conjugate and (b) disjunctive movements of the eyes.

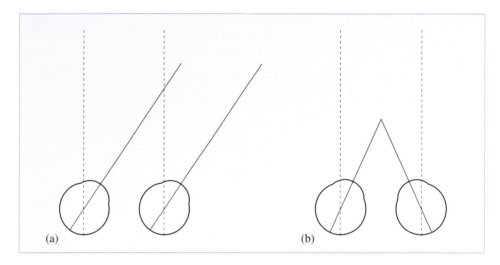

Clearly there is a limit to the **convergence** that can be achieved in disjunctive movements – you can test this for yourself by holding a pencil vertically in front of you and moving it towards your nose while looking at it. The point at which you see two pencils is the near point of convergence – for most people it is between 5 and 10 cm from the front of the eyes and is considerably closer than the nearest point at which you can focus clearly.

We can separate out three different types of conjugate eye movements, all of which serve different purposes.

3.6.2 Conjugate movements

Saccades

The word 'saccade' comes from the French verb, *saccader*, meaning to jerk or flick. **Saccades** are the most common form of eye movement and are rapid and abrupt conjugate jumps made by the eyes as they move their point of fixation from one location to another. These are some of the fastest movements in the body with a speed of up to $1000°\,\text{s}^{-1}$. They can be small ($< 0.05°$) or large ($20°$) and occur as the eye moves from one point of fixation to another.

What is their purpose? It is believed that they are used to explore the visual field and to place images of selected visual details on the fovea, where the ability to resolve detail is best. They are very important for reading, and in the examination of stationary scenes. Figure 3.29 confirms that the movements are predetermined – it shows how the saccades observed when looking at a scene vary according to the task given to the observer.

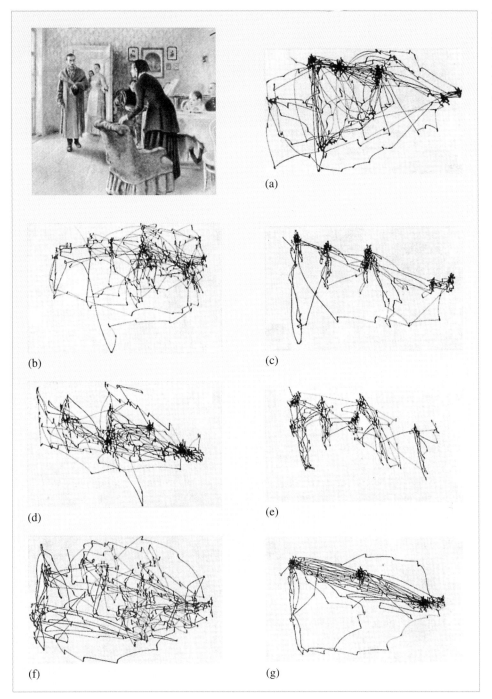

Figure 3.29   Seven records of eye movements by the same subject viewing the same picture but under different instructions in each case. Each line represents a saccade and each record lasted for 3 minutes. Both eyes were used. For (a) the subject was allowed free examination of the picture.
For the subsequent recording sessions the subject was asked to:
(b) estimate the material circumstances of the people in the picture;
(c) give the ages of the people;
(d) surmise what the family had been doing before the arrival of the 'unexpected visitor';
(e) remember the clothes worn by the people;
(f) remember the positions of the people and objects in the room;
(g) estimate how long the unexpected visitor had been away from the family.

One might wonder why we do not see a blurred image during the saccade itself. In fact why are we not even aware that our own eyes are making these movements? This lack of awareness of any blur during saccadic eye movements is referred to as **saccadic omission** and is thought to be due to visual masking. The stimuli perceived just before and after the eye movement mask the blur caused by the movement. This has been confirmed by experiment.

Pursuit movements

These are exactly what their name suggests – movements that allow the eyes to pursue moving objects. They are almost completely automatic movements and are much slower and smoother than saccades; the maximum speed is about $40° \, s^{-1}$. **Pursuit movements** allow better perception of the moving object because they maintain the image of the object stationary on the fovea, where spatial resolution is best (Figure 3.30).

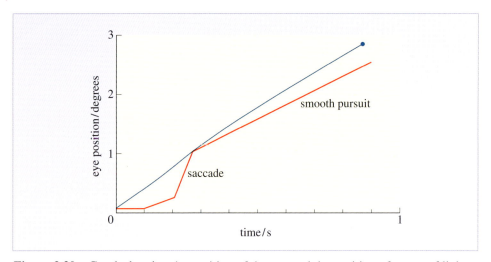

Figure 3.30 Graph showing the position of the eye and the position of a spot of light that acts as the target the eye is following. The target (blue spot) starts to move at time zero. Note that at first the eye does not move (there is a reaction time). Then it starts a slow smooth pursuit movement, but soon the observer realizes that the target is moving ahead of the eye so a corrective saccade is made. After that a smooth pursuit movement is made which follows the spot of light. Note that this entire process only covers an angular distance of about three degrees and takes about one second.

Vestibulo-ocular movements

Vestibulo-ocular movements enable the eyes to fix on a point despite movement of the head.

○ Keeping your head still, move this book from side to side at a gradually increasing speed until you are no longer able to read the words.

Now keeping the book in front of your eyes move your head from side to side so that the angular speed is about the same. What happens now? Can you read the book?

Can you explain this phenomenon?

● You should have found that you can continue reading when your head is moving but not if you hold your head still and move the book. This is because, when you move your head, your eyes are moving so as to compensate exactly for the movements of your head.

It is possible to compensate so accurately for head movements because the acceleration of the head is detected by the vestibular system in the inner ear. The vestibulo-ocular reflex then causes the eye muscles to operate in such a way as to maintain the gaze.

You can see the vestibulo-ocular reflex in action if you ask someone to hold their gaze on your nose and then to move their head from side to side. Observe the movement of the eyes relative to the head. You should note that, as the head moves to the right, the eyes remain looking in the same direction and are therefore being rotated to the left relative to their sockets. The movements are smooth, rapid and exact.

Movements in response to the vestibulo-ocular reflex can be as fast as $300°\,s^{-1}$. The vestibulo-ocular system will be discussed further in Block 7.

3.6.3 Disjunctive movements

Our eyes face forward but they do not both capture exactly the same image because they are a little apart from one another. The left eye can see slightly further to the left and the right eye slightly further to the right. In order to see objects clearly, each eye is moved independently by its muscles until the object of interest is situated in the same place on each retina. The eyes must converge more to focus on near objects and converge less to focus on objects at a distance.

These movements tend to be slower than the convergent movements described already and happen automatically so that the two retinal images are aligned. If the movements do not happen correctly the result is **diplopia**, or double vision. Diplopia is symptomatic of any failure of the eyes' musculature (or their associated neurons and brain regions). One cause of diplopia is the consumption of large quantities of alcohol!

These kinds of movement are most significant when the focus of the eyes is changed from a distant to a close object. This is a reflex movement – the lines of sight will change so as to converge onto the object without your being aware of it.

Box 3.6 Eyeblinks

Eyeblinks are not a movement of the eyeball itself but a movement of the eyelid that covers the eye. If you watch someone else's eyes you should note that they blink approximately 15 times per minute – more often if they are startled, anxious, excited or stressed in any way, and less often if they are carrying out a task that is visually demanding, such as reading this course material!

The interesting thing about eyeblinks is that, like saccades, we are not aware of an interruption of vision while we are blinking. A blink takes a few hundred milliseconds. If the lighting in the room is turned out for that long we are aware of it, but we are not aware of blinks. This is thought to be due to an inhibitory signal that suppresses the visual stimuli during the blink.

When reading, the blink frequency drops to about 5 per minute. If you can manage to observe someone reading without their being aware of it, you will notice that blinks do not occur randomly but are most likely to happen when the reader is at the end of a line, a sentence or a paragraph. This is a further indication that they happen when the intake of visual information is at its lowest.

3.7 Adaptation to different levels of illumination

The human eye can cope with an amazing range of luminance. Figure 3.31 shows typical values.

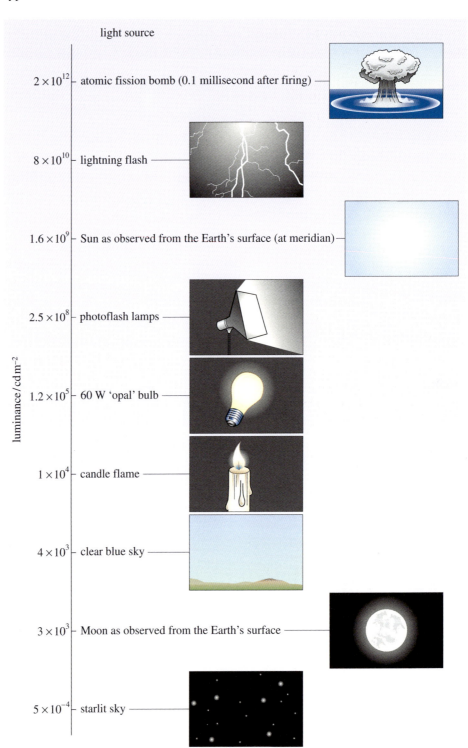

Figure 3.31 The approximate luminance values of different sources. The eye can be damaged by looking at the first three without protection but the remaining sources, which cover a large range, can all be observed. (The units used are photometric units and need not concern us as we are interested only in the relative values. Note also that the scale is not linear.)

3.7.1 Alterations in pupil size

The simplest and most obvious way in which the eye copes with this range is by altering the size of the pupil. The pupil can alter its diameter from a minimum of 2 mm to a maximum of 8 mm. This four-fold change in the diameter means that there can be a sixteen-fold change in the area and therefore in the amount of light entering the eye. This is not enough to account for the eye's ability to cover all the lighting situations described in Figure 3.31, but it is none the less useful, not least because it is a fast response. Figure 3.32 shows this effect, which is known as the direct light reflex (Section 3.2). Note the dotted line representing the left eye, which was not exposed to the bright light but which none the less altered in size via the consensual light reflex. Note also the gradual increase in the pupil diameter after prolonged exposure – this is due to adaptation of the retina to the light level, which will be considered in Section 3.7.2.

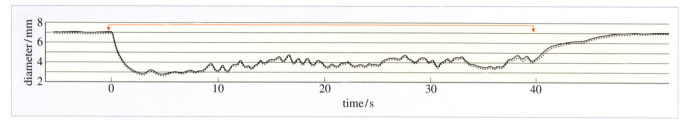

Figure 3.32   The effect of exposure of the right eye to a bright light. The solid line records the diameter of the right pupil against time, the dotted line represents the diameter of the left pupil. The plot shows the immediate effect of exposing the right eye to a bright light (at the arrow). After several seconds the pupillary oscillations can be seen. When the light is turned off the pupils dilate and the oscillation disappears.

The average extent of the pupillary reflex depends on the light level in the manner shown in Figure 3.33. On this diagram note the points corresponding to the threshold for colour (cone) vision and for rod vision.

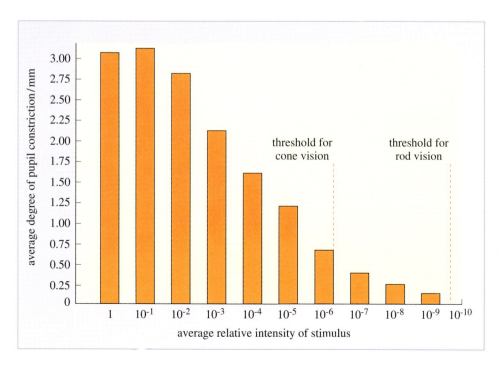

Figure 3.33   The average extent of pupillary reflexes elicited by light stimuli of different intensities and duration of one second. The heights of the columns indicate the average pupil degree of constriction while the horizontal axis shows the relative intensities of the stimuli.

Box 3.7 Red eye

We have all seen indoor photographs in which the people looking towards the camera all have red eyes. The reason for this is that the pupils are dilated because the ambient light levels are low. When the camera flash is initiated, it causes a rapid contraction of the pupil. However the time taken for this to occur (0.25–0.5 s) is more than the time taken for the camera to take the picture. The result is that red light, reflected from the **choroid** at the back of the eye, is observed on the image.

Figure 3.34   A typical indoor flash photo showing the 'red-eye' effect.

○ Why do we not observe this effect in normal situations where the light level is low?

● When we look into someone else's eyes our head always blocks the light that would travel straight through to the choroid and back again. In amateur flash photography the source of light (the flash) and the detector (the film) are in line and the light from the flash can reach the choroid and return to the film.

Many cameras now have a built-in system for reducing the red eye effect. They send a preliminary burst of light to the eye about 0.75 s before the exposure of the film. This causes the pupils to constrict. Then a second flash occurs at the moment the film is exposed. Alternatively, a professional photographer will use a flash linked to the shutter but held at an angle some distance from the camera so that the light is not reflected from the choroid to the film.

Other than the obvious reduction or increase in the number of photons reaching the retina, what effect does the alteration of pupil size have? There are three effects we can consider.

1 Depth of field

The depth of field, at a fixed focus, is the distance (in depth) through which a point object can be moved without appearing blurred. It depends on the inability of the retinal system to discern any blurring when the size of the blur circle is below a certain limit. (The threshold is usually taken to be equivalent to having the lens 0.15 D away from the correct value.) Figure 3.35 shows how pupil size affects the size of the blur circle – clearly a smaller pupil leads to less blurring. Table 3.2 shows the depth of field for different pupil sizes.

If you are over forty years of age you will almost certainly have noticed that reading is easier in bright light. This is because as you age your eye is less able to accommodate for close-up vision (see Section 3.3.4). If the object you are viewing is well illuminated, your pupil is smaller, your depth of field is increased and you can

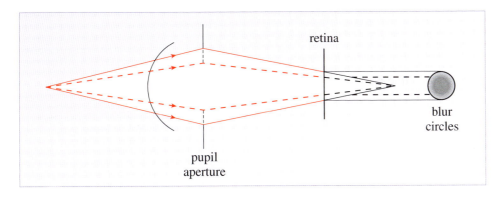

Figure 3.35 The effect of pupil size on the blur circle formed by an out-of-focus image on the retina. For a small pupil size (dashed lines) the blur circle is reduced and objects over a larger distance can be seen sharply; the depth of field is increased.

Table 3.2 The variation in depth of field for different pupil sizes.

Pupil diameter/mm	Distance of focused object from eye/m	Minimum distance (in metres) at which the object is seen sharply	Maximum distance (in metres) at which the object is seen sharply
4	infinity	3.5	infinity
4	1	0.8	1.4
2	infinity	2.3	infinity
2	1	0.7	1.8

view closer objects. You may also have seen short-sighted people half-close their eyes in order to see objects in the distance – they are making use of this increase in depth of field by narrowing their 'pupils' even more using their eyelids.

2 Aberrations

Anyone who is an enthusiastic photographer will know that if you want a good camera you have to pay a lot of money for a good lens. This is because lenses that focus all the light to a point are very difficult to make. Any lens that does not do this is said to have aberrations. We can distinguish two important types of aberrations.

Spherical aberrations

In Section 3.3.2 it was suggested that a lens with spherical surfaces would focus light to a point. In fact this is only true for rays which pass through the lens close to the centre. Off-centre rays tend to focus to a different point. Figure 3.36 (overleaf) shows this effect.* Since many lenses are made by grinding glass to a spherical surface, these **spherical aberrations** can be important in optical instruments.

The larger the pupil size the worse the possible effects of spherical aberrations. However they are never a serious problem in the normal eye, even for large pupil diameters. The reason for this is that the cornea is not spherical but is flatter at the periphery than in the centre and is therefore closer to the ideal shape for avoiding spherical aberrations.

* You can also observe a similar effect (in two dimensions) by looking into a cup of tea! The light reflected from the inner surfaces of the cylindrical cup is not all reflected back to the same point. Instead it forms a 'cusp'. To avoid this effect, headlight reflectors are parabolic and not spherical.

Figure 3.36 Spherical aberrations arise because, if the lens surfaces are spherical, light passing through the outer portion of the lens is not focused to the same point as light passing through the lens closer to the axis.

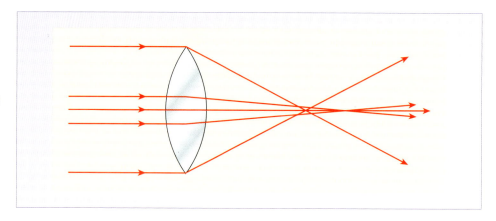

Chromatic aberrations

It was also mentioned in Section 3.3.2 that the refractive index of a material depends on the material and on the wavelength. It is generally larger for wavelengths at the blue end of the spectrum than it is for red wavelengths. The focal length of the lens/cornea system is therefore different for different wavelengths. The consequence of this is that if, with the eye focused at infinity, yellow light is focused on the retina, then blue light will be focused in front of the retina and red light behind (see Figure 3.37). Calculations show that the dioptric effect of these **chromatic aberrations** is quite considerable (equivalent to about 1.5 D). However most of the time we are not aware of it – perhaps because the extremes of the spectrum where the refractive error is large are also the regions of the spectrum where the detectors are less efficient.

There is now considerable evidence to suggest that chromatic aberrations are part of the mechanism that allows the eye to accommodate (see Section 3.3.3).

Figure 3.37 Chromatic aberrations arise because the refractive indices in glass for different wavelengths of light are slightly different. The index for blue light is larger than the index for red light so the blue light is refracted more. The effect has been exaggerated here for clarity.

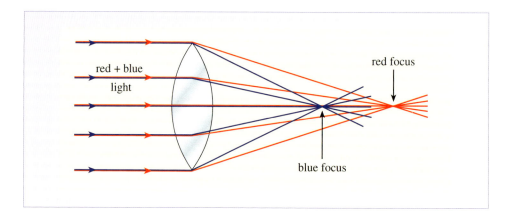

3 Diffraction

Both depth of field and aberrations are worse when the pupil is dilated. But there is one effect that is worse when the pupil diameter is small, and that is diffraction. Box 3.4 explained the concept of diffraction in the context of an array of slits (or molecules in the case of the cornea). It is also possible to have diffraction as light passes through a single slit or a circular aperture. The patterns that this gives rise to are described in Box 3.8.

Box 3.8 Diffraction at a single slit

A diffraction grating gives rise to a diffraction pattern of light and dark regions because of constructive and destructive interference between wavefronts emerging from different slits. The diffraction pattern due to a single aperture arises because of interference between wavefronts leaving different points of the aperture. The mathematics need not concern us here – the important thing is to look at the shape and size of the pattern produced.

Figure 3.38 shows the pattern that would be obtained on a screen some distance from a single slit of width w. (The observed light intensity corresponds to the square of the amplitude.) The light is clearly spread out and the amount of spreading is inversely proportional to the slit width – a narrow slit gives a wider diffraction pattern than a broad slit.

In this example we have considered a slit so that the diffraction is only noticeable in one dimension. However the analysis can be extended to two dimensions. One well-known example is the pattern obtained from a circular aperture. This was first explained by Airy in 1835 and is now known as an Airy disc (see Figure 3.39 overleaf).

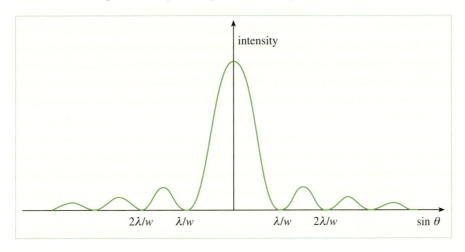

Figure 3.38   Diffraction pattern from a single slit. Note that the first minimum occurs at $\sin \theta = \lambda/w$. This means that the width of the central maximum is inversely proportional to the slit width.

The pupil is a circular aperture so the diffraction pattern obtained is an **Airy disc**. Figure 3.39 (overleaf) shows the intensity pattern in an Airy disc. Note that the first minimum occurs at a radius of $1.22 f\lambda/D$ where f is the focal length of the lens, λ the wavelength and D the diameter of the pupil.

○ Calculate the radius of the first dark ring on the retina for (a) a pupil of diameter 2 mm and (b) a pupil of diameter 8 mm. Take the wavelength of the light to be 600 nm (6×10^{-7} m) and the focal length of the eye to be approximately 17 mm.

● (a) For a 2 mm pupil:

$$\text{Radius of first dark ring} = \frac{1.22 \times \left(6 \times 10^{-7}\,\text{m}\right) \times \left(1.7 \times 10^{-2}\,\text{m}\right)}{2 \times 10^{-3}\,\text{m}} = 6.2 \times 10^{-6}\,\text{m}$$

(b) For an 8 mm pupil diameter the answer would be four times smaller, 1.6×10^{-6} m.

The significance of these sizes will be discussed when we look at visual acuity in Section 3.8. Table 3.3 (overleaf) summarizes the effects of changing pupil diameter.

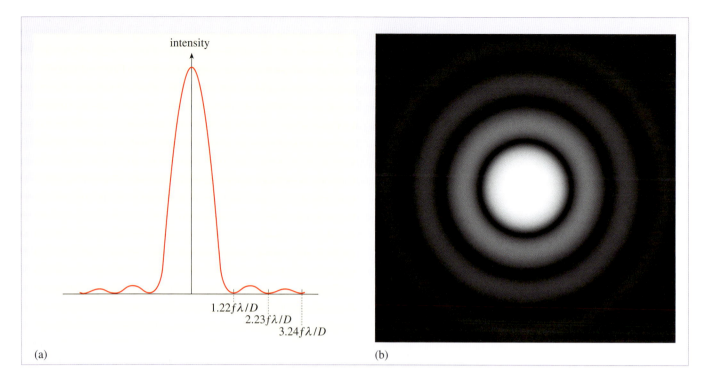

Figure 3.39 The diffraction pattern for a circular aperture. (a) Intensity plotted against radius; (b) the diffuse pattern of concentric rings around a central disc known as an Airy disc.

Table 3.3 The effect of changing pupil diameter.

Causes of blurring	Effect of increasing pupil diameter	Effect of decreasing pupil diameter
reduction in depth of field	increases	decreases
spherical aberrations	increases	decreases
chromatic aberrations	increases slightly	decreases slightly
diffraction	decreases	increases

It was stated earlier that the change in pupil diameter is not enough to account for the eye's ability to cover a very wide range of luminance. To see how the eye does cope we need to look at the way that the retina is affected by the amount of light reaching it.

3.7.2 Changes in the retina

The brightest sunlight encountered in normal life is more than 10^{12} times brighter than the dimmest starlight. How can the visual system possibly function efficiently over such a huge range?

Fortunately, the full range of brightness never occurs simultaneously; the brightest object in a given scene is unlikely to be more than a few thousand times brighter than the dimmest. The visual system is thus able to function in a rather ingenious way. The processes that encode brightness are only sensitive to a relatively limited range of variation, but the visual system automatically adjusts its sensitivity to bring the prevailing lighting conditions into this range. This process of adjusting to the prevailing lighting conditions, called **dark adaptation**, is rather like sliding a small magnifying glass up and down a very long ruler to make accurate measurements at whatever point they are needed. Dark adaptation does not take place instantaneously as can be seen from Figure 3.40.

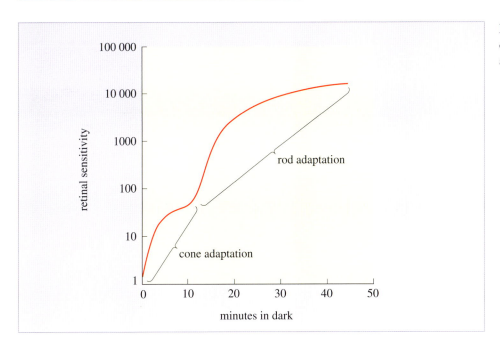

Figure 3.40   Dark adaptation, demonstrating the relation of cone adaptation to rod adaptation.

○ Describe your own experience of moving from bright sunlight into a darkened room.

● At first it is difficult to see anything at all but gradually you can discern shapes and then more detail.

○ How long does it take to get a ten-fold increase in sensitivity when you first go from light to dark? (Use Figure 3.40.)

● A ten-fold increase in sensitivity occurs in 1–2 minutes. (Note that the scale is logarithmic and is therefore not constant over the range. This allows very large changes to be represented on the one graph.)

In fact after 30 minutes the sensitivity of the retina has increased about 10 000-fold. The changes that bring about the increased sensitivity occur in the photopigment found in the membrane stacks of the outer segment of the rods and cones. You may recall from Section 3.4 that the pigment is bleached by bright light. The amount of photosensitive chemical available to respond to photons is thus considerably reduced when you first move from bright to dark conditions. However, regeneration of the pigment occurs relatively rapidly in the cones, thus enabling them to respond to the low level of illumination. Full regeneration of the rod pigment rhodopsin takes longer but finally achieves far greater sensitivity.

3.7.3 Neural adaptation

There is a third mechanism that contributes to dark adaptation, and like the pupillary response it is very rapid. It takes place in the retina, and is termed **neural adaptation**. It is essentially due to the dynamic reorganization of ganglion cell receptive fields so that the inhibitory sub-regions are less powerful. Inhibitory sub-regions are important in resolving spatial detail but inhibition inevitably reduces sensitivity. When photons are scarce, the loss of inhibition maintains sensitivity at the expense of spatial resolution. Although this adaptation is only a fewfold, it does, as mentioned above, occur almost instantaneously.

Question 3.10

When the focus of the eyes changes from a distant to a close object there are three changes that take place in the eyes. What are they and why do they happen?

Question 3.11

In view of what you have learned about rods and cones, why do you think driving is particularly difficult at dusk?

Summary of Sections 3.6 and 3.7

In principle, the eyes can be rotated about all three axes but normally only rotate about the vertical and the transverse horizontal axes. Movements can be conjugate or disjunctive but in both cases the movements of the two eyes are equal and symmetric.

Conjugate movements fall into three different types:

1   saccades, which are very rapid, abrupt, conjugate movements;

2   pursuit movements, which are slower, smooth movements that allow the eye to follow a moving target;

3   vestibulo-ocular movements, which are controlled by the vestibular system in the inner ear and which compensate for movements of the head.

Disjunctive, or vergence, movements are used to avoid double vision and take place when the focus of the eye changes from far to near or vice versa.

The eye can adapt to variations in illumination by a factor of 10^{12} by altering the pupil size and by changes in the retina.

The amount of light entering the eye can only be reduced by a factor of 16 by decreasing the size of the pupil over the whole range of movement. Reducing the size of the pupil increases the depth of field of the eye and also reduces effects due to aberrations. However diffraction effects are more likely to be noticeable with a small pupil.

Most of the remaining adaptation occurs in the retina where rhodopsin is slowly regenerated when the light level is reduced.

There is also a neural response to changing light levels which is fast but less effective.

3.8 Visual acuity

Whether or not we can see things clearly is something that concerns all of us. And those of us with poor eyesight spend a lot of money at the opticians trying to improve our **visual acuity**! This section looks at the way visual acuity can be measured and at the factors affecting it.

3.8.1 What is meant by visual acuity?

Visual acuity can be defined in several different ways. Table 3.4 describes some of the types of acuity that have been described and gives examples of the kinds of images that might be used to test acuity.

Table 3.4

Acuity	Description	Example of objects used to test acuity
detection acuity	the ability to detect a small object in the visual field	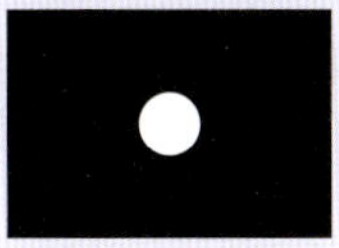
resolution acuity	the ability to detect a separation between two discrete elements of a pattern	
recognition acuity	the ability to recognize the object (e.g. letters, as in the familiar Snellen test, consisting of rows of letters of decreasing size)	
vernier (or localization) acuity	the ability to detect whether two lines, laid end to end, are continuous or offset	
dynamic acuity	the ability to detect and locate a moving target	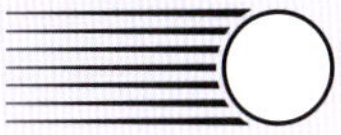

Of these measures of acuity the best known is the Snellen test which tests **recognition acuity** and which is the test that is normally used in clinical practice.

The vernier test is also a very interesting one because it turns out that the **vernier acuity** of the human eye is extremely high compared with the other measures of acuity.

However, from our point of view we can start to get a good measure of the eye's acuity if we consider **resolution acuity**.

3.8.2 Estimation of visual acuity

Let's take a very simple approach to the measurement of resolution acuity. Look at the two lines drawn in Figure 3.41. Prop this book up vertically on a table or shelf at eye height in a well-lit room and, starting from a distance of about one metre, walk backwards while looking at the lines.

What happened? Close up, you could see that there were two lines on the page. But as you walked backwards away from the page the lines gradually merged into one another until you could no longer see that they were two lines.

Let us consider what is happening in your eye when you look at the two lines. An image of those lines is formed on the retina. This is shown in a simplified way in Figure 3.42 (overleaf). In this diagram the retina is shown as an array of receptive elements.

In Figure 3.42a the images of the lines fall on two receptors which are separated by another receptor. In this case the brain can distinguish the two lines. In Figure 3.42b

Figure 3.41 These two lines are 0.5 mm wide and 0.5 mm apart.

Figure 3.42   An image of the two lines is formed on the retina. The ray diagram (not to scale) has been simplified and the retina is shown as an array of receptive elements. (a) If the images of the two lines are sufficiently far apart for there to be a receptive element between them then the lines will be distinguished.
(b) If the images of the lines fall on two adjacent receptors then only a continuous image will be seen.

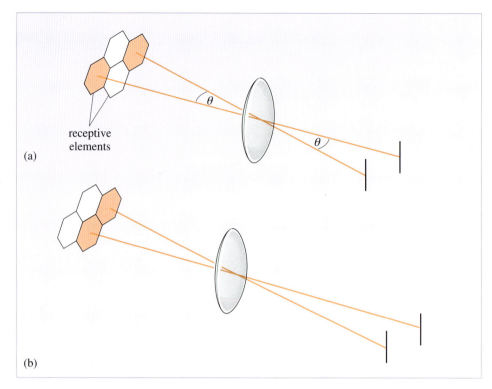

the images of the lines fall on two adjacent receptors so that the image seen is exactly the same as the image of a continuous thick black line, i.e. the two lines cannot be distinguished.

The lines used in the two diagrams in Figure 3.42 have exactly the same spacing; however they are at different distances from the eye. The deciding factor is the spacing on the retina, and this is determined by the angle between the two lines at the lens. This **visual angle** is shown as θ in Figure 3.42a. The size of this angle for two objects that can just be distinguished is a measure of the resolution acuity of the eye.

You can use the lines in Figure 3.41 to carry out a simple experiment to estimate the minimum angle required for you to be able to distinguish the two lines. To do this experiment you should prop the book up vertically as before. If you wear glasses for distance vision then use them for this experiment. As before, walk backwards away from the page until the lines merge into one another. Measure your distance from the page at this point (let us call the distance L). You will probably find that this does not happen suddenly so it will be difficult to get an exact measurement, but you should try to get a value to the nearest half metre.

○   The lines in Figure 3.41 are 0.5 mm apart. Use your estimate of the value of L to find a value for the angle θ in degrees.

●   Because the distance, d, between the two lines is very small compared with the measured distance L, we can say that $\tan \theta \approx d/L$ (Figure 3.43). d is given as 0.5 mm (5×10^{-4} m) and let us suppose that you have measured L as 2 m. Then $\tan \theta = (5 \times 10^{-4}$ m$)/2$ m $= 2.50 \times 10^{-4}$, and, using a calculator, $\theta = 1.4 \times 10^{-2}$ degrees.

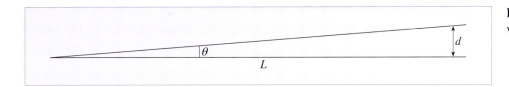

Figure 3.43   Since the angle θ is very small, $\tan\theta \approx d/L$.

Your answer is probably very much less than one degree. When considering such angles it is customary to use **minutes** (often referred to as '**minutes of arc**' to avoid confusion with time). One minute of arc is 1/60 of a degree so you can convert from degrees to minutes by multiplying by 60. For the answer above the value obtained is 0.86 minutes. Write *your* value for θ in the box here for reference later.

Value for θ for your eyes = …… degrees = …… minutes of arc

3.8.3 Definition of visual acuity

The standard definition of visual acuity, in terms of the angle θ that you have measured, is:

$$\text{visual acuity} = 1/\theta \qquad (3.11)$$

where θ is measured in minutes of arc.

On this scale the average person has a resolution acuity of about 1, that is to say the angle θ that they would measure is 1 minute of arc.

So how does your visual acuity compare with the average? If you have good eyesight and carried out the test under good conditions you may well find that your value of θ is smaller than 1 minute, so that your visual acuity is nearer to 2. On the other hand, if conditions were less than ideal, you may find that your visual acuity is less than 1. In that case, before you dash off to the opticians, remember that this is only a very rough and ready test! There are several other factors that determine visual acuity, one of the most important being the lighting conditions. This will be discussed in more detail in Section 3.8.5.

3.8.4 The predicted value of visual acuity

In Section 3.8.1 it was suggested that visual acuity is dependent on the way in which the receptors on the retina are activated. If there is an inactive (or less active) sensor between two strongly activated sensors then the eye can detect the presence of a gap between two lines (Figure 3.42). You are now in a position to check whether or not this is a plausible explanation for the visual acuity values observed. If you refer back to Figure 3.42 you can see that the limit of visual acuity must also be given by the equation:

$$\tan\theta \geq \frac{\text{diameter of one receptor on the retina}}{\text{distance between the retina and the centre of the lens/cornea system}} \qquad (3.12)$$

since the minimum possible separation of the images of the lines must be the diameter of a receptor.

○   At the centre of the eye (the fovea) the receptors are cones of diameter approximately 1.5 μm. The distance between the centre of the lens system and the retina is usually taken to be about 17 mm. What value does this give for the visual acuity?

●   Using the figures given, the minimum value for θ is 0.3 minutes (or about 20 seconds). This would correspond to a visual acuity value of just over 3.

Given that this is a minimum value and that the best measured values of visual acuity are approximately 2, this calculation would tend to support the suggested explanation for the excellent visual acuity of the human eye.

Box 3.8 20/20 Vision?

You will probably have heard people talk of having 20/20 or 20/30 vision. This is a way of comparing visual acuity. The ratio is:

$$\frac{\text{the farthest distance at which the subject can read a line of letters}}{\text{the farthest distance at which an average person with normal vision can read the same letters}}$$

The measurement is made at 20 ft, hence the 20 in the ratio. Accordingly a ratio of 20/15 means that the subject can read at 20 feet what the average person can only read at 15 – good acuity. A ratio of 20/30 means that the subject can only read at 20 feet what an average person can read at 30 feet – poor acuity.

Normal vision on this scale corresponds to an acuity of 1 on the scale we used above in equation 3.11. On a metric scale the distance used is 6 m, hence normal vision is 6/6, etc.

It is a little frightening to realize that the UK driving test only requires us to have 6/14 vision!

3.8.5 The contrast sensitivity function

This simple test of resolution acuity gives a rough description of the performance of the eye. A far more general description of the limits of human visual performance is provided by the **contrast sensitivity function** (**CSF**). This is a function that gives a measure of the sensitivity of the eye at different spatial frequencies. Before we consider the contrast sensitivity function we need to look at a more general concept used in imaging – the modulation transfer function.

Ideally an image should be an exact replica (probably magnified or reduced) of the object and should therefore contain all the spatial frequencies that were present in the object. In practice this is not the case – with any imaging system, including the eye, there is almost always some degradation that occurs during the actual imaging process. The amount of degradation is crucial in determining the quality of the imaging system.

If there is a lot of detail (i.e. high spatial frequencies) in an object then obviously it is desirable for those high spatial frequencies to be transmitted by the imaging system to the final image. The transfer of spatial frequencies is described by a transfer function. The simplest transfer function to consider is the **modulation transfer function** (**mtf**).

Modulation is the term used to describe the variation in shade of an object or image. Consider the pattern shown in Figure 3.44. This is a simple sinusoidal variation of shade from dark grey to almost white with a spatial frequency $1/\theta$ (Section 2.4). The variation of intensity can be represented by the waveform shown in Figure 3.44b, which demonstrates that there is only the one spatial frequency present in this case. The modulation at this spatial frequency is calculated using the expression:

$$M = (I_{max} - I_{min})/(I_{max} + I_{min}) \tag{3.13}$$

where I_{max} represents the maximum intensity and I_{min} the minimum intensity.

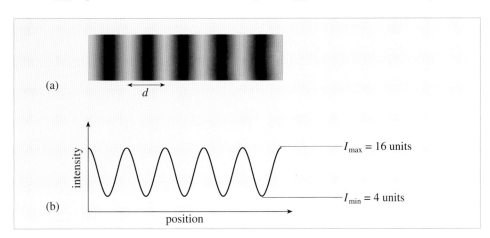

(a)

d

(b)

intensity

position

$I_{max} = 16$ units

$I_{min} = 4$ units

Figure 3.44 (a) A sinusoidal intensity pattern. (b) The intensity plotted against position. The modulation is given by $M = (I_{max} - I_{min})/(I_{max} + I_{min})$. In this case $M = (16 - 4)/(16 + 4) = 0.6$.

If the maximum intensity were white and the minimum black then the modulation would be 1; for all other variations, including the one shown in Figure 3.44 the modulation is less than 1.

The modulation transfer function describes how well the imaging system transfers this modulation from object to image. At each spatial frequency the mtf is defined by:

$$\text{mtf} = \frac{\text{modulation in image at a particular spatial frequency}}{\text{modulation in object at the same spatial frequency}} \tag{3.14}$$

If the values of the mtf are plotted against spatial frequency the resulting curve gives an indication of the performance of the imaging system.

A perfect imaging system would transfer all the information perfectly, so the modulation transfer function would be 1 at all frequencies. In practice this is not possible (except for zero spatial frequency, i.e. a uniform field) and the mtf normally decreases with increasing spatial frequency. Even a 'perfect' lens has an mtf that reduces as spatial frequency increases. In the case of the eye we can use the concept of modulation transfer function to examine the properties of the image produced on the retina by the dioptric system of the eye. Figure 3.45 (overleaf) shows the typical mtf of the human eye.

For the overall performance of the eye, things are a little more complicated because the ability of the retina and brain to detect different spatial frequencies also comes into play. The properties of receptive fields in the retina mean that the eye is actually less sensitive at spatial frequencies of zero than it is at a few cycles per degree.

Here the term **contrast** is used instead of the term modulation and the usual way to represent the sensitivity of the whole system is to plot the contrast sensitivity function.

Figure 3.45   The modulation transfer function for the human eye with a 2 mm pupil compared with that for a perfect optical system with the same aperture. Note that there is a progressive loss of modulation as the spatial frequency increases, confirming that higher spatial frequencies are harder to transfer to the retina.

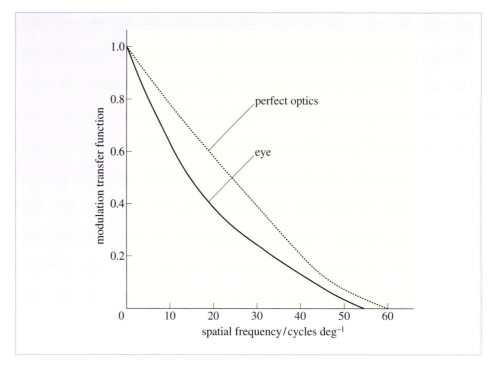

Contrast threshold is the minimum contrast at which an observer can reliably detect the presence of a grating. **Contrast sensitivity** is just the reciprocal of threshold, so that a low threshold indicates high sensitivity, and a high threshold indicates low sensitivity. As shown in Figure 3.46, the CSF is just a plot of sensitivity at each spatial frequency, measured using sine-wave gratings. Note the additional (dotted) line which shows what the CSF might be expected to be if it depended only on the mtf of the dioptric system.

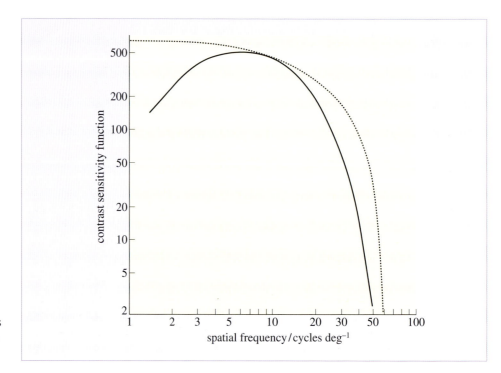

Figure 3.46   The contrast sensitivity function gives a measure of the variation in sensitivity with spatial frequency. The solid line is the measured CSF, the dotted one is the CSF calculated assuming that the mtf is the only factor affecting CSF. The pupil diameter was 2.5 mm.

In practice, the CSF only describes vision at near threshold levels, and it does not apply to suprathreshold, clearly visible, stimuli. Despite this limitation, the CSF has proved extremely useful in providing a very rich description of visual performance. For example, it incorporates a measure of acuity, as the finest grating (i.e. highest spatial frequency) that can be detected at the maximum possible contrast (100%). The typical frequency of 50–60 cycles per degree (i.e. 1 cycle per minute of arc) agrees well with conventional measures. As shown in Figure 3.46, it also reveals that we are not most sensitive to a uniform field (i.e. zero cycles per degree) but to a 4–8 cycle per degree grating. This fits in with the known properties of retinal ganglion cells (Section 3.4); their receptive fields contain antagonistic sub-regions which make them relatively insensitive to uniform illumination. Note too that at the optimal frequency, our vision is exquisitely sensitive, with a contrast threshold of only 0.2%.

Figure 3.47 provides a quick way for you to confirm that your own CSF follows approximately the same form as that measured more rigorously in Figure 3.46. Hold Figure 3.47 in front of you at arm's length and look at it for several seconds. You should be able to visualise a curve between the area where you can see the bands and that where you can't. This curve represents your CSF.

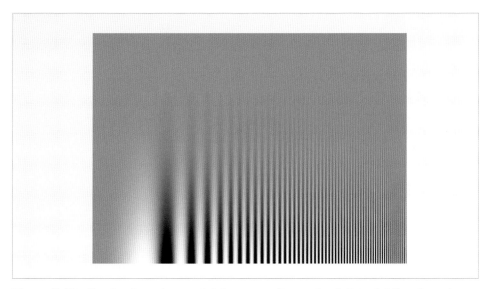

Figure 3.47 Bands of varying spatial frequency (increasing left to right) and varying contrast (modulation) (increasing top to bottom). The subjective contour separating the visible and invisible lines has the approximate shape of the contrast sensitivity function.

The CSF also provides a way to describe how acuity, and other aspects of visual performance, depend upon such factors as retinal position and the overall level of illumination. Figure 3.48 (overleaf) shows how the CSF changes when the stimulus is presented to the retinal periphery, rather than the fovea, and Figure 3.49 (overleaf) allows you to check this for your own eyes.

Figure 3.50 (overleaf) shows how the CSF changes when the overall level of illumination is decreased. Our acuity falls dramatically, as the finest resolvable grating shifts to a much lower spatial frequency, the whole curve shifting to the left and downwards. The CSF thus provides a precise way of confirming and measuring the familiar impression that things are more blurred and harder to see unless we look at them directly in reasonably bright conditions.

Figure 3.48   The relative visual acuity at different positions on the retina. All values are expressed as proportions of the foveal acuity. Note the blind spot where the optic nerve enters the eye.

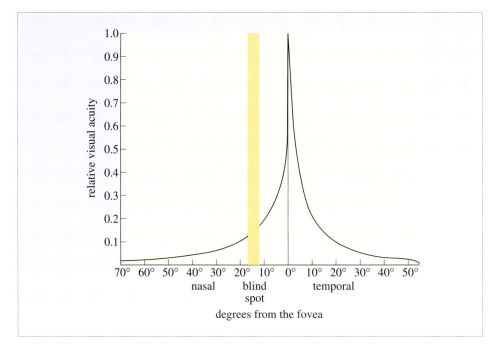

Figure 3.49   A demonstration that acuity decreases rapidly away from the fovea. The chart is designed so that, if you fix your eyes on the spot at the centre, all the letters appear equally legible.

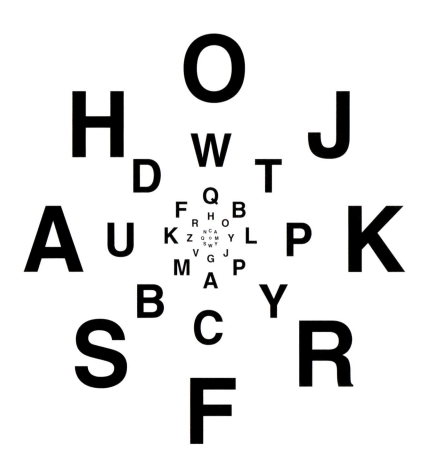

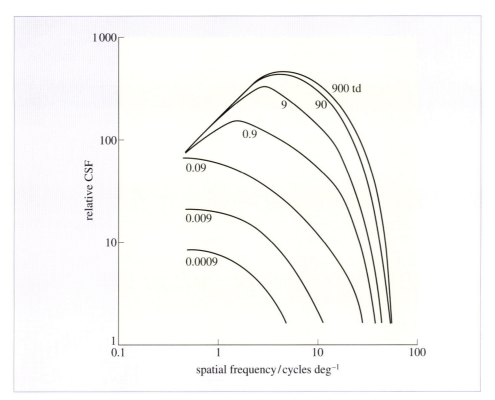

Figure 3.50 CSF curves at seven different retinal illuminance levels. (The units used, trolands (td), are a measure of retinal illuminance. The highest value corresponds to the greatest illuminance.) In these tests the size of the pupil was kept constant by using an artificial pupil of 2 mm diameter.

One factor in the dramatic fall-off in acuity under scotopic conditions and for peripheral viewing could be the distribution of cones on the retina. Figure 3.51 shows the relative numbers of rods and cones in the retina. The distribution of cones matches the acuity curve of Figure 3.48. But this hypothesis completely disregards the other type of receptor present in the retina, the rods. In the central foveal region there are no rods but their density increases towards the edges of the retina. This would suggest that at night, when the rods are more important than the cones, peripheral acuity would be better. To a certain extent this is true, but it is still far

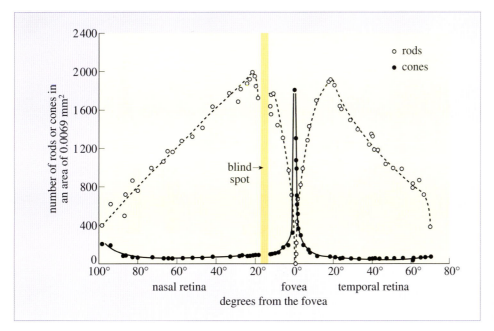

Figure 3.51 The distribution of rods and cones in the eye. Cones are concentrated in the fovea and rods in the outer areas.

below the standard of photopic foveal acuity. There are clearly other very important factors to be taken into consideration, including the receptive fields of retinal ganglion cells. You may recall from Section 3.4 that receptive fields are smaller in the fovea than in the retinal periphery, and since the retinal ganglion cells form the output from the retina this will clearly be an important factor in accounting for the loss of acuity with peripheral viewing. Similarly, the receptive fields derived from rods are generally larger than those derived from cones, and this will be important in the loss of acuity at low illuminance. The relationship between the CSF and receptive fields will be described in more detail in Section 4.

Question 3.12

There is a popular claim that the Great Wall of China is the only artefact on Earth that is visible from space with the naked eye. The wall is approximately 6 m wide. A spaceship that is orbiting the Earth needs to be at least 200 km above the Earth to avoid the drag caused by the Earth's atmosphere. If the visual acuity required for observation of the Great Wall is determined by the same criteria as for two lines, what do you think of the credibility of this claim?

Question 3.13

Try testing your own visual acuity again with the two lines in Figure 3.41 but this time have the lines horizontal in one case and oblique in the other. Compare these results with the original test with the lines vertical.

Question 3.14

In Section 3.3 the eye was compared to a camera and several differences between the two were mentioned. However one aspect that was not discussed was the acuity and sensitivity of the detector. In a camera the film has uniform acuity and sensitivity across its surface. How do these factors vary for the eye? How does the human eye optimize the situation?

Question 3.15

The 'best' pupil diameter is often taken to be 2 mm and an artificial pupil of this size is often used for testing aspects of vision. Can you explain why 2 mm might be considered to be a good size? What happens if the pupil size is (a) smaller, or (b) larger than this?

Question 3.16

Why is the CSF a good measure of visual performance?

Summary of Section 3.8

There are several different measures of visual acuity. The best known of these is the Snellen test which is used in clinical practice and which measures recognition acuity.

Resolution acuity is determined by measuring the angle (θ) at the eye between two objects that can just be distinguished. It is defined by:

visual acuity = $1/\theta$ where θ is measured in minutes.

The average value is 1 and this is in good agreement with the value obtained if it assumed that objects can only be distinguished when they do not excite adjacent cones on the retina.

The quality of any imaging system, including the eye, can be expressed in terms of the modulation transfer function (mtf) of the system. Modulation is defined as:

$$M = (I_{max} - I_{min})/(I_{max} + I_{min})$$

and the mtf as:

$$\text{mtf} = \frac{\text{modulation in image at a particular spatial frequency}}{\text{modulation in object at the same spatial frequency}}$$

A perfect imaging system would transfer all the information perfectly, so the modulation transfer function would be 1 for all spatial frequencies. In practice this is not possible (except for zero spatial frequency) and the mtf normally decreases with increasing spatial frequency.

A better measure of the acuity of the visual system is given by the contrast sensitivity function (CSF). This combines the modulation transfer function of the dioptrics with the sensitivity of the retinal system to different spatial frequencies.

Visual acuity is strongly affected by the angle of vision (eccentricity) and the level of illuminance. Increasing eccentricity and lower light levels both lead to lower visual acuity. Both effects can be explained by looking at the relationship between the receptors and the pathways to the brain.

From eye to brain

4

4.1 Introduction

This section completes the coverage of the initial stages of visual processing. It begins by briefly introducing the anatomical and functional pathways from the retina to the cortex, highlighting the way that different aspects of visual processing are performed by different types of specialized mechanism and in different parts of the brain. It goes on to describe the different processes that underpin specific aspects of vision, such as spatial form, lightness and brightness, colour, motion and some aspects of depth perception. The section ends by considering how the rather fragmentary results of this initial processing might be combined to provide the coherent representations that are needed by the complex processes of visual perception.

Although the emphasis here is primarily upon processes that deal with the retinal image, it is important to keep in mind that vision involves much more than this. The retinal image is a small, upside-down, two-dimensional pattern of light that jerks about as we move our eyes. Yet we visually experience a stable, three-dimensional, correctly-oriented world of meaningful objects. We cannot fully explain this experience if we consider only the image. This point should be brought home by Figure 4.1. Is this a cube? Think for a moment about the properties of the object being depicted. It is solid, consisting of *square* surfaces joined together at *right angles* to each other. Now look again at Figure 4.1, which is an *image* of a cube. It is a flat pattern of lines forming three regions, without squares or right angles. How, from such unpromising beginnings, does your visual system effortlessly and immediately work out that it represents a cube? The answer must surely be that vision involves more than just a description of the image, for that would just be a description of some strangely shaped regions. Rather, the visual system must do something much more difficult; it must work out what object might have produced the image. We perceive the result of this process, not the image that begins it.

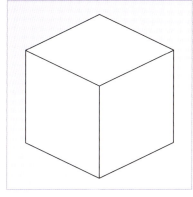

Figure 4.1 An image of a cube. Compare its properties with those of a real cube.

In the light of this, it is often convenient to think of vision as two different types of process. The first is concerned with describing the image and is often called **bottom-up** processing because it starts with the image and builds upwards toward higher levels of processing. The second is concerned with making sense of this description and is often called **top-down** processing because it starts with high-level knowledge about the world and uses this at lower levels of processing. This section deals with bottom-up processing, and Section 5 deals with top-down processing. Although the processes you will encounter in Section 4 are essentially descriptive, you should not expect them simply to produce a faithful copy of the image. Rather, their role is to extract from the image those features that are particularly informative about the objects that produced the image, and that thus make subsequent, top-down processing easier.

4.2 Pathways from the retina to the cortex

The aim of this section is to provide you with an overview of the relevant visual anatomy. You should not try to remember all the detail at this stage. Instead, you can refer back to the figures as the different structures are discussed in the text. Figure 4.2 (overleaf) summarizes the relevant retinal anatomy.

Figure 4.2   The basic anatomy of the retina. (This is a simplified version of Figure 3.20.)

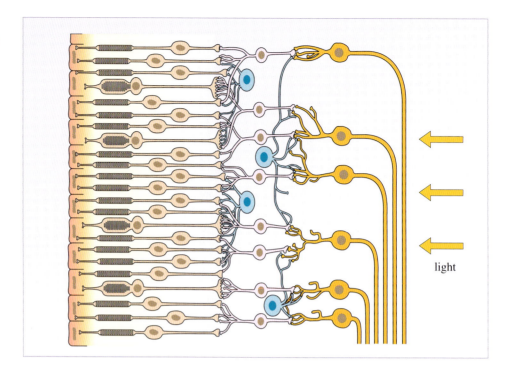

light

First, small, densely-packed photoreceptors transduce luminance into a high-resolution 'neural map' of the image. Second, there is a complex network of excitatory and inhibitory interneurons allowing rich interactions between neighbouring parts of the retina. This network is responsible for most of the processing that occurs in the retina. Finally, retinal ganglion cells carry the output of this retinal processing from the eye to the brain and are the first stage in the neural pathway in which action potentials, rather than graded potentials, occur. The second half of Chapter 10 of the Reader, *The retina* by Jim Bowmaker, which you encountered in Section 3.4, explains in more detail the interconnections between receptors and ganglion cells.

The axons of the retinal ganglion cells travel across the surface of the retina, leave the eye together at the blind spot, and become the individual fibres of the optic nerve. Figure 4.3 shows the main pathway of the optic nerve from the eye to the brain, the **geniculostriate pathway**. The optic nerves from the two eyes first meet at the **optic chiasm**, where half the fibres cross over to the other side of the brain and the other half remain on the same side. As shown in Figure 4.3, this arrangement ensures that the two 'neural maps' of the same region of visual space project to the same side of the brain. The fibres continue to the first synapse in the pathway, at the **lateral geniculate nucleus** (**LGN**) in the thalamus.

The LGN is divided into six layers and the cells in each layer are monocularly driven, i.e. each layer receives input from just one of the eyes. Each layer is **retinotopically mapped**, i.e. the receptive fields of adjacent LGN cells cover adjacent regions of the image. Finally, different layers receive input from different types of retinal ganglion cell and project to different types of cell in the visual cortex. The different layers may therefore be involved in processing different kinds of information, although the function of the LGN is not yet fully understood. LGN cells have very similar properties to the retinal cells that drive them, suggesting that the LGN does not perform much actual processing, and, indeed, it may simply act as

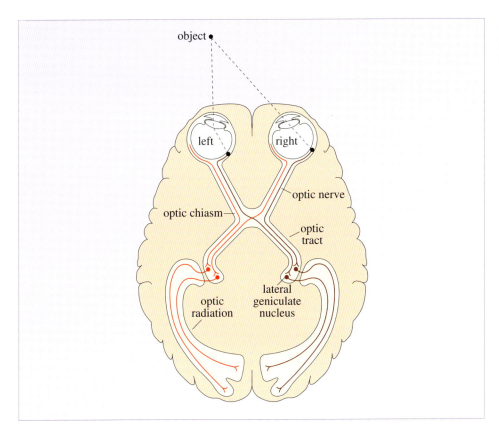

Figure 4.3   The main visual pathway from the retina to the cortex.

an anatomical relay station, sorting out the different fibres in the optic nerve into different cell types and different functions. However, there is also a very large downward projection from the cortex that may yet prove important.

○   What role might the cortical input to the LGN play?

●   One possibility is that it provides a route by which later processes can select the information that is passed on to them.

You should now read the first part of Chapter 12 of the Reader, *From retina to cortex* by Andrew Derrington, to the end of Section 3, which provides details about the projections to and from the various layers of the LGN. This chapter covers a very complex part of the visual system, so you shouldn't worry if you find it hard going on first reading.

Fibres from the LGN project via the **optic radiation** to the visual cortex at the back of the brain in the occipital lobe (Figure 4.4 overleaf). This region is variously called the **primary visual cortex** (because it is the first cortical stage of visual processing), the **striate cortex** (because its prominent layers give it a characteristic striped appearance in anatomical preparations), **area 17** (from Brodmann's numerical classification of cortical areas) or, in primates like human beings, **V1**. These are all equivalent names for the same area. Like the LGN, V1 is divided into layers, running parallel to the cortical surface, and the layers are retinotopically mapped. The map is distorted in that the cortical representation of the fovea is much larger than its physical size on the retina. It is also rather coarse-grained, in that there are typically many thousands of cortical cells representing each small region of the retina. These cells are organized roughly into columns, at right angles to the layers,

Figure 4.4 Some of the important cortical areas concerned with visual processing. MT: medial temporal area; MST: medial superiortemporal area. The top view represents the brain pulled apart at the sulcus running down the temporal lobe to reveal the tissue that normally forms the wall of the sulcus, and so is hidden from view. The bottom view represents the brain with the occipital lobe pulled back to expose the sulci between the occipital and parietal lobes at the top, and the occipital and temporal lobes at the bottom.

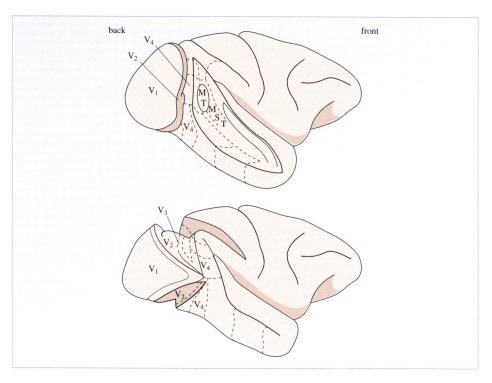

and the set of columns that processes each small region of the retina is called a **hypercolumn**. The cells within each hypercolumn are also organized at a finer level into discrete anatomical structures, each of which contains cells specialized for one particular aspect of visual processing. In effect, you can think of the hypercolumn as the neural machinery needed to represent all the potentially useful features of one tiny part of the image – one piece of a complex neural jigsaw that must be solved by later processes. The relevant functional anatomy of the hypercolumn will emerge in Section 4.3, as we consider cortical functions in more detail.

Activity 4.1 The optic chiasm and retinotopic mapping

Both the function of the optic chiasm and the concept of retinotopic mapping are further explained on *The Senses* CD-ROM and you may find it useful to review this material before moving on to Section 4.3. Further details are given in the Block 4 *Study File*.

V1 is just the first cortical stage of visual processing. From here, visual information is passed on to over 30 currently identified cortical regions, each concerned either exclusively with vision, or requiring an important contribution from vision. The main areas involved in early visual processing are shown in Figure 4.4. In these initial stages of the complex network, cortical processing seems to be organized by **visual sub-modality**, with separate regions dealing with, for example, colour or motion information. We will consider these separate aspects in more detail in Section 4.3.

The later stages of cortical processing, however, seem to be organized very differently. Neuropsychological evidence suggests that separate regions may be concerned with different visual tasks. Some stroke patients who have been unfortunate enough to suffer localized lesions to the visual system, for example, seem unable to recognize a visually-presented stimulus such as a hammer, while remaining perfectly able to mime its function. Other patients, with differently located lesions, may have normal vision except that they are unable to recognize faces. Such cases suggest that our normally rich conscious visual impressions of the world may arise from interactions between many functionally-specialized modules, each responsible for just one aspect of visual processing.

This modular organization is not surprising when we consider that vision consists of many different tasks, requiring different, and sometimes incompatible, information. Compare, for example, the problem of using vision to recognize an object with that of using vision to pick it up. To recognize something, we need information about the *relationships* between features like edges and corners but, since recognition must work from many different viewpoints, we need to ignore the absolute spatial positions of these features. On the other hand, if we need to *pick up* the object, the *absolute positions* of the features are crucially important. It may well be then, that like the 'what' and 'where' systems that you encountered in the auditory system, the visual system needs completely separate modules for recognizing objects and for guiding visual actions. This view is certainly supported by an elegant experiment based on the Titchener circles, shown in Figure 4.5.

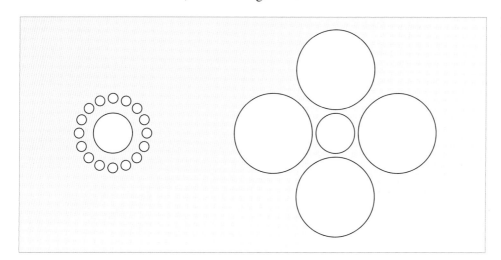

Figure 4.5   The Titchener circles. The central circles are the same physical size.

The central circles in both versions of the figure are physically identical in size, but the central circle surrounded by large circles looks smaller than the one surrounded by small circles. When people are asked to judge the perceived size of the central circles, they are consistently taken in by this illusory effect. When, in another test, they try to pick up circular targets of various sizes, they automatically adjust their initial finger positions very accurately to the size of the targets. The interesting finding is that, despite consciously perceiving the Titchener circles to be of *different* sizes, people none the less adjust their fingers to the *same* size when trying to pick them up. This raises the intriguing possibility that our conscious representations of visual objects are separate and different from the visual representations that we use to guide our actions; we may see the world wrongly but we still deal with it correctly.

In addition to these neuropsychological and experimental studies, it has recently become possible to investigate functional visual modules more systematically, using brain imaging techniques.

You should now read Chapter 13 of the Reader, *Functional imaging of the human visual system* by Krish Singh, which explains these techniques and their role in investigating the functional modularity of the visual cortex.

4.3 Visual sub-modalities

In this section we will consider the processes that deal with spatial structure, luminance intensity, colour and relative distance. In the early stages of cortical processing, the visual system deals separately with each of these visual sub-modalities. However, it is important to remember that, though we are forced here to deal with each aspect in sequence, the visual system can and does deal with them simultaneously and in parallel, through a set of specialized mechanisms, each dealing with one aspect of the visual stimulus.

4.3.1 The coding of spatial structure

Retinal processing

One of the main functions of retinal processing is to reduce the neural 'image' registered by the photoreceptors to a neural 'line drawing' registered by the retinal ganglion cells. This is achieved by the network of retinal interneurons, described in Chapter 10 of the Reader, *The retina*, which is arranged so that the receptive field of each ganglion cell consists of two concentric sub-regions of roughly equal strength. Consequently, when its receptive field is uniformly illuminated, the excitatory and inhibitory influences cancel each other out, and the ganglion cell does not change its response from its normal resting firing rate. However, when the illumination is not uniform, for example when a luminance edge falls anywhere within the receptive field, the excitatory and inhibitory influences do not balance and the ganglion cell responds, either by decreasing or increasing its firing rate. Several demonstrations on *The Senses* CD-ROM, which you first encountered in Section 3.4, remind you of the details of this important point.

Figure 4.6 illustrates the pattern of firing produced in an array of ON-centre ganglion cells by a typical luminance edge. On each side of the edge, all the photoreceptors respond to exactly the same degree. Relatively few ganglion cells respond, but these few responses are enough to signal that a change in illumination exists, where it is, and the magnitude of the change. Thus, ganglion cell receptive fields greatly reduce the data that need to be transmitted to the brain, without losing any of the potentially useful information; if you know where the changes are and how big they are, you can easily reconstruct the original image. Figure 4.7 (overleaf) demonstrates that the visual system can perform precisely this kind of reconstruction. This stimulus looks like a conventional light-dark edge, but both sides of the figure actually have the same luminance. The central region of the stimulus is designed to produce the same pattern of response in ganglion cells as would be produced by a real edge; and given a response that usually signals the size and position of an edge, the visual system automatically delivers the conscious impression of an edge.

Initial stages of cortical processing

The retina passes on only the most informative features of the image. The initial stages of cortical processing begin to extract and describe the most useful aspects of these features. Cells in area V1 can be classified into types according to the characteristics of their receptive fields. The most important types are shown in Figure 4.8 (overleaf).

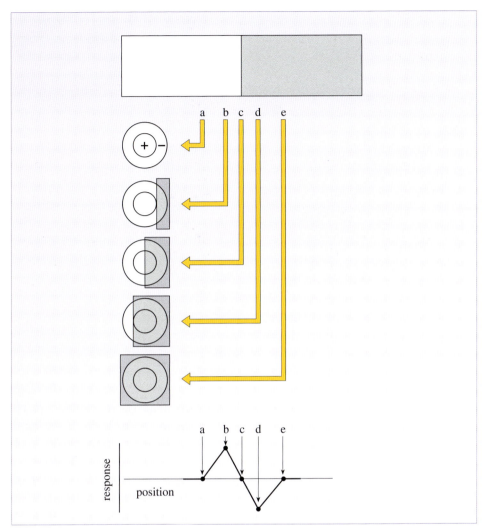

Figure 4.6 The pattern of response across an array of ON-centre retinal ganglion cells. The top part of the figure shows a light/dark edge. The middle part shows the positions of receptive fields of a few representative cells (a–e) relative to the edge; the left hand column shows the pattern of illumination of each receptive field. The bottom of the figure shows the responses of the cells. (a) The receptive field is uniformly illuminated, the centre and surround responses are in balance, so the cell does not respond. (b) All of the excitatory centre but only part of inhibitory surround is illuminated; the centre response dominates the surround response so the cell increases its response rate. (c) Half the excitatory sub-region and half the inhibitory sub-region are illuminated, the centre and surround responses are in balance, so the cell does not respond. Note that this cell has its receptive field directly under the edge. (d) Part of the inhibitory sub-region is illuminated, the surround response dominates so the cell decreases its response. (e) The receptive field is in uniform darkness, the centre and surround responses balance, so the cell does not respond.

Figure 4.7   The Craik-O'Brien-Cornsweet illusion. The two outer edges of the figure appear to have different lightnesses, even though both are physically the same. The luminance profile of the stimulus is shown below the figure. Note that the illusion depends to some extent upon the viewing distance; you may find it most effective to view the figure from a short distance.

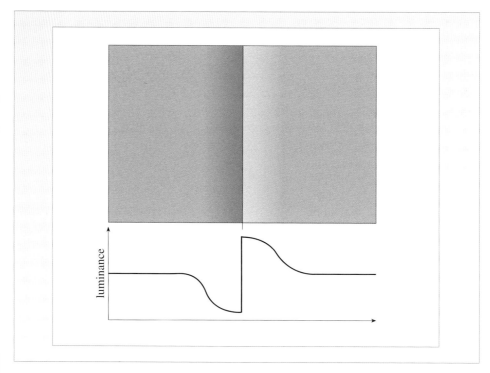

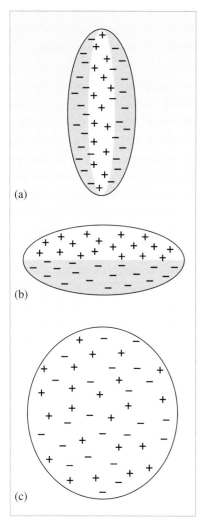

The receptive fields of **simple cells** (Figure 4.8a and b), like those of retinal ganglion cells, consist of distinct antagonistic sub-regions. However, the receptive fields are typically elongated so that, unlike retinal ganglion cells, simple cells are selective for stimulus orientation; each simple cell responds best to a line (Figure 4.8a) or an edge (Figure 4.8b), not only at a specific position but also at a particular orientation in the image. Simple cells probably derive their receptive fields by combining the input from a number of retinal ganglion cells at adjacent positions on the retina.

The receptive fields of **complex cells** (Figure 4.8c) also show excitatory and inhibitory influences, but they are not arranged in discrete antagonistic sub-regions. Despite this, complex cells respond best to a line or edge at a particular orientation but, unlike simple cells, they are less specific for stimulus position; they respond to an appropriately oriented stimulus anywhere within the receptive field. One final important class of cells, called **end-stopped**, or **hypercomplex**, **cells** have receptive fields like those of complex cells, but unlike them, respond only if an appropriately oriented stimulus ends somewhere within the receptive field.

Early accounts of cortical function emphasized the role of individual cells. Simple cells, for example, were proposed as **feature detectors**; the response of a given simple cell was thought to signal the presence of a line or edge at a particular position and orientation in the image. Complex cells were thought to collect the input from several appropriately positioned simple cells, so that they could detect the presence of an oriented feature anywhere within a region of the image.

Figure 4.8   Typical receptive fields of cells in the primary visual cortex (V1).
(a) Simple cell: the receptive field is divided into discrete antagonistic sub-regions. This cell would respond best to a vertical bar. (b) Simple cell. This cell would respond best to a horizontal edge. (c) Complex cell: a larger receptive field without discrete sub-regions. Despite this, the cell would still respond best to an appropriately oriented bar or edge.

Subsequently, however, it was realized that individual cells are not selective enough in their responses to fulfil these roles. A simple cell, for example, will respond to a range of stimulus orientations, so that it cannot unambiguously detect a single orientation. Consequently, modern theories focus upon the pattern of response across whole populations of cortical cells.

Within the hypercolumn, remember, cells are organized anatomically into columns, and intracellular recordings in the monkey reveal that all the cells within each column have the same preferred orientation. Moreover, preferred orientation varies smoothly from column to column so that, within each hypercolumn, all orientations are represented in a sort of 'pinwheel' arrangement. Other aspects of the stimulus, such as its colour and direction of motion, are also represented within the hypercolumn, so that we can imagine a complex pattern of cortical response across many thousands of cells that signals the presence, size, orientation, colour and direction of movement of lines or edges within each small region of the image.

You should now read the rest of Chapter 12 of the Reader, *From retina to cortex*. Section 4 provides more detail about the properties of the various types of cell and about how their receptive fields may be derived. Section 4.3 summarizes the functional and anatomical arrangement of the hypercolumn.

In trying to understand in detail how the pattern of response within the hypercolumn might signal spatial structure, vision scientists make great use of the contrast sensitivity function (CSF) introduced in Section 3.8.5. Indeed, the sine-wave grating, with which the CSF is measured, can fairly be described as a kind of 'psychophysicists' microelectrode', allowing the properties of visual mechanisms to be inferred from the measurable responses of the whole human observer. Not only does the overall shape of the CSF tell us about the types of retinal and cortical mechanisms that we have already described, but relatively simple techniques allow us to work out how individual cells might respond to any given stimulus. It thus becomes possible to predict quite accurately what the pattern of response in cortical hypercolumns might be, and even to suggest how this pattern might be used in describing the spatial structure of the image.

You should now read Chapter 14 of the Reader, *Spatial vision* by Tim Meese, which summarizes this approach, and which concludes this section on the coding of spatial structure.

4.3.2 The coding of luminance intensity

Section 3.7 discussed the problems caused by the wide range of variation in natural illumination and how the visual system adjusts to the prevailing lighting conditions by a process known as dark adaptation. In this section, we turn to another aspect of this problem. Once we have adapted to the prevailing light, how does the visual system encode the changes in **luminance** within a particular image? This is not a straightforward problem, as is illustrated in Figure 4.9 (overleaf). Most objects reflect, rather than emit light and the amount of light that a surface reflects depends upon the amount of light striking the surface (the **illuminance**) and upon the proportion of this light that the surface reflects (the reflectance of the surface, see Section 2.2). If the visual system simply measured the amount of light at any point in the image, it would inevitably confound these two factors and you would not be able to distinguish between a well lit, unreflective surface and a badly lit, reflective one.

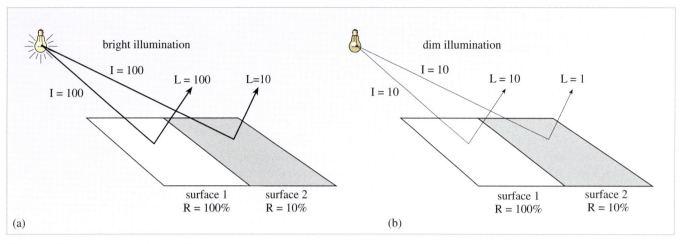

Figure 4.9   The physical basis of lightness constancy.

In fact, of course, we are not fooled in this way. **Lightness constancy** refers to the fact that a grey (i.e. neutrally reflective) surface continues to look grey over a huge range of lighting conditions; it does not look white (strongly reflective) when well lit, or black (weakly reflective) when badly lit. Somehow, the visual system solves the problem of recovering information from the image about an unvarying and often characteristic property of objects (surface reflectance) despite potentially confusing variations in illuminance. Indeed, the single physical dimension of image luminance really gives rise to two separate perceptual dimensions: the perceptual impression of surface reflectance is called **lightness**, while the perceptual impression of illuminance is called **brightness**. Thus it makes sense, for example, to talk of a light surface that is badly lit.

How is lightness constancy achieved? Look again at Figure 4.9. Although the luminance of each surface varies with the illumination, there is something about this situation that always stays the same.

○ Can you see what it is?

● The *ratio* of the light from the two surfaces always remains the same; the surface on the left always reflects 10 times as much light as the surface on the right, because it is 10 times as reflective. Notice that this ratio is, in effect, the contrast of the edge ($(L_{max} - L_{min})/(L_{max} + L_{min})$).

How might the visual system make use of this constant feature? One solution would be to compare the image luminances at adjacent points in the image. Most of these comparisons would lead to a value of 1, since at most places in the image, neighbouring luminances are the same. But across the edge, the ratio of the luminances would give a measure of the **relative reflectance** of the surfaces (i.e. 10).

○ What sort of familiar visual mechanism might compare the luminances at neighbouring image points?

● The antagonistic sub-regions of the receptive fields of retinal ganglion cells and cortical simple cells mean that these cells, in effect, compare the luminance at adjacent image regions.

Of course, the comparison needs to be made as a ratio, rather than, say, a simple difference. But retinal and cortical cells do seem to perform the appropriate arithmetic because their responses are proportional to the contrast across an edge (i.e. a ratio), rather than to the difference in luminance across the edge.

One less desirable consequence of calculating ratios in order to extract information about relative reflectance is that information about the *absolute* luminance is lost. There are many other ways of recovering this information (for instance from the processes that control dark adaptation), but it is certainly true that our perception of absolute luminance is actually rather poor. For example, television images often seem to contain black regions that are much darker than the screen appears when the set is turned off. But a television screen cannot actually be physically darker than when it is turned off, because cathode rays can only brighten the screen; dark regions are just brightened less than light regions. You see apparent blacks because the visual system is relatively insensitive to *absolute* luminance, and more concerned with *relative* luminance. Relatively dark regions appear black in the television image because they are much darker than the lighter regions of the image and because, within a given context, the visual system interprets 'relatively dark' as 'black', and 'relatively light' as 'white'.

This phenomenon is called **lightness contrast** and is just an inevitable consequence of lightness constancy. Both phenomena are illustrated in Figure 4.10. The top strip appears uniform while the bottom strip appears lighter at one side than the other. Physically the reverse is true; the top strip varies gradually in physical luminance from one side to the other and the bottom strip is physically constant throughout its length. You can verify this by cutting a strip-sized hole in a large piece of paper and then viewing each strip in isolation from the background. Both percepts are to be expected in a system that signals relative reflectance by taking the ratio of the

Figure 4.10   Lightness constancy and lightness contrast. The background luminance varies continuously from one side of the figure to the other. The luminance of the top strip varies continuously from one side to the other; this strip appears uniform because it has the same contrast at every point (lightness constancy). The bottom strip is physically uniform but appears brighter at one end because its contrast varies from one point to the next (lightness contrast).

luminance across edges. Here the background varies in such a way that the ratio between the top strip and its immediate background is everywhere the same, and we see this constant ratio as lightness constancy. But the ratio between the bottom strip and the background varies, and we see this varying ratio as lightness contrast. Of course, lightness constancy is a sensible solution to the real problem of varying illumination in the natural world, whereas lightness contrast is just an artificial illusion that is only likely to be a problem in psychology laboratories.

These examples support the idea that the responses of retinal and/or cortical cells are important in signalling relative lightness. But these responses are not the *only* factor that determines lightness perception.

○   What everyday luminance phenomenon cannot be explained by retinal/cortical responses signalling relative lightness?

●   A shadow will generally produce a luminance edge in the image and, consequently, a response in retinal and cortical cells. But the visual system correctly interprets a shadow as a change in illumination, rather than a change in surface reflectance.

In practice, simple retinal and cortical mechanisms simplify the problems caused by varying illumination and they explain several perceptual phenomena. But, as we shall see in Section 4.4, they cannot provide a complete solution to the problem of interpreting image luminance.

4.3.3 The coding of colour

Although colour vision may seem technical, it is actually one of the simplest of visual sub-modalities and, consequently, the best understood. You already know a great deal about the retinal processes involved from Chapters 8, 10 and 11 of the Reader. In the first stage of neural processing, at the photoreceptors, human colour vision is **trichromatic**; three different types of cone analyse image luminance into broad bands of wavelengths: long (red), medium (green) and short (blue). At the next stage, in the network of retinal interneurons and the retinal ganglion cells, human colour vision begins to show **colour opponency**; information about the separate bands of wavelengths is combined in a variety of ways, most notably by cells that compare different bands by responding positively to one but negatively to the others.

Both trichromacy and opponency seem to be sensible adaptations to the pattern of wavelengths typical of natural light and to the kind of colour information likely to be found in natural images. Moreover, between them they can explain many perceptual colour phenomena. Trichromacy, for example, can explain many of the rules of colour mixing, and the colour anomalies sometimes found in human vision. Opponency, on the other hand, can explain our perception of complementary colours and coloured afterimages. However, although it may thus seem that most of colour vision can be explained by retinal processes, the first stages of the visual cortex also play an important role.

Information about colour is relayed to the LGN, where colour-selective and non-colour-selective cells are segregated into separate layers. These layers project to specific regions of the hypercolumns in V1, called '**blobs**' because of their appearance when stained for microscopic examination. Cells in the blobs, in turn, project to a distinct area of the temporal cortex called V4 (Figure 4.4), which seems to be specifically concerned with colour. Colour information seems to be gradually refined along this path from retina to cortex, so that in V4 individual cells respond only to very narrow bands of wavelengths, or specific colours. The processing involved can be understood as a simple extension of lightness perception.

We saw in Section 4.3.2 that lightness perception involves recovering information about the reflectance of surfaces. Colour perception is a refinement of this process, and involves recovering information about the spectral reflectance of surfaces. Rather than recovering a single number that gives the total amount of light that the surface reflects, we need to recover a spectrum of numbers that gives the amount of light reflected at each wavelength. The advantages of this refinement are obvious if we imagine coloured and black and white photographs of the same scene, or compare our vision at twilight, when cones are relatively unresponsive, with normal daytime vision. In the black and white photo, or at twilight, we can generally distinguish objects from their backgrounds though many objects appear identical because they have the same overall reflectance; all cats are grey at night. But in the colour photo, or in daylight, these objects are immediately distinguishable from each other because they reflect different patterns of wavelengths.

We also saw in Section 4.3.2, that the problem in lightness perception is to extract information about surface reflectance despite variations in the overall illumination, and that this problem could be solved simply by taking the ratio of luminances at edges in the image. The additional problem for colour vision is to extract information about spectral reflectance despite not only variations in overall illumination, but also variations in the wavelengths of the illumination. Full sunlight provides a very different pattern of illuminating wavelengths than, for example, sunlight filtered through trees. Consequently, the pattern of wavelengths reflected by surfaces is very different in these two lighting conditions. But, despite this, we exhibit **colour constancy**; we generally see similar surfaces as more or less the same colour, irrespective of the colour of the light shining upon them.

Colour constancy can be based on the familiar trick of taking the ratio of luminances at edges, because this calculation in effect cancels out any effects of illumination. But now this ratio must be calculated separately for each band of wavelengths. In effect, colour constancy is largely lightness constancy performed separately within each of the three cone systems.

You should now read Chapter 15 of the Reader, *Colour in context: contrast and constancy* by Anya Hurlbert, which discusses colour constancy in more detail, and which concludes this section on the coding of colour.

4.3.4 Measuring relative depth: motion parallax and binocular stereopsis

In Block 1 and Section 1 of this block, we saw that there are many potential depth cues allowing the recovery of depth and distance from two-dimensional images. Many of these cues are fairly high level, requiring prior knowledge, or assumptions about the scene. Such cues give rise to *inferred* depth; when we look at a simple black-and-white perspective line drawing, for example, we know that one object is meant to be further away than another, but we do not have a compelling three-dimensional subjective experience of depth. However, in addition to these high level cues, there are also simpler cues that can be measured directly and that give valuable information about depth. For example, the degree of convergence needed to bring an object into register in both eyes, and the amount of accommodation required to flatten the lens to focus on an object both provide potential measures of object distance.

They are sometimes dismissed as being useful only at short distances up to about 1 m but, since this is about arm's length, it is exactly the range in which depth is most important in allowing us to manipulate objects.

Two other simple depth cues both depend on comparing more than one image of the same scene from different viewpoints. There are two ways in which this can be done. First, we can compare images taken at different times as we move about the world using **motion parallax**. Second we can compare simultaneous views from the two eyes, using **binocular stereopsis**. Unlike our experience of inferred depth, each of these cues is sufficient by itself to produce an immediate and compelling subjective experience of depth. In that sense, they are the most important cues in determining our normal conscious impression of visual depth.

Motion vision

Motion is a visual sub-modality in its own right. Although there are no true motion sensors in the human retina, retinal cells do provide the first stage of a specialized motion-processing system. The retinal cells you encountered in Section 4.3.1 are primarily concerned with recovering spatial structure from the image by picking out those regions where the luminance changes from one position to the next (i.e. luminance edges). Other retinal cells have complementary properties and seem to be concerned with recovering the temporal structure of the image by signalling changes in luminance from one instant to the next. These two types of cell are anatomically distinct and project to separate layers in the LGN (see Chapter 12 of the Reader, *From retina to cortex*, for a little more detail). At the cortex, in V1, responses of the two types of cell are combined and many cells in this region can be described as simple motion sensors. These cells are directionally selective, in that they respond more to one direction of motion than any other. Directionally-selective cells in V1 project to a specialized motion area, the medial temporal area (MT, sometimes called V5), where **direction selectivity** is further refined, and then on to the medial superiortemporal area (MST) where the responses of many motion cells are combined to produce cells selective for specific patterns of retinal motion (Figure 4.4).

Motion is particularly useful in providing information about the three-dimensional structure of the world because, as you know from looking out through the side-windows of a moving car or train, our own movements produce characteristic patterns of motion in the image. The most relevant aspect of these patterns is that distant objects move more slowly in the image than do nearby objects. This difference in retinal speeds is an example of motion parallax, an important, directly measurable depth cue. Its value is illustrated in Figure 4.11. People generally perceive this figure as a rectangular object tilted in depth so that the top edge is further away than the bottom edge. But this requires an assumption about the shape of the object – after all, it might be a trapezoid that is not slanted at all. But if this were a real object and we moved relative to it, no such assumption would be required. If the top edge were further away, then it would move more slowly in the image than the bottom edge. In fact, the ratio of the retinal speeds of the top and the bottom edges gives the relative distances of the edges. Without motion, we can *infer* depth based on assumptions about the shape of things. With motion, we can simply *measure* relative depth.

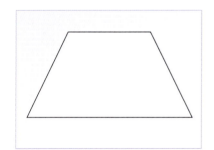

Figure 4.11   The usefulness of motion parallax. This figure might represent a square tilted in depth or, for example, a trapezoid in the plane of the page. Movement of the observer would disambiguate the stimulus because the resulting image motion depends on the distance of the stimulus; if the stimulus is a tilted square, the bottom of the figure would move faster in the image than the top.

You should now read Chapter 16 of the Reader, *Motion vision* by Mike Harris, which provides details about how visual motion is encoded and about how retinal motion may be used in timing complex everyday tasks, such as knowing when and how hard to brake when driving.

Binocular stereopsis

Stereopsis literally means solid vision. As its name implies, binocular stereopsis recovers depth information, or solidity, by comparing the images in the two eyes. This is made possible by our forward facing eyes, which ensure that each eye sees more or less the same region of the world, and by the optic chiasm (Section 4.2, and Figure 4.3), which ensures that information from both images about the same region of visual space projects to the same side of the brain. Retinal and LGN cells receive input from only one eye, so the first stage at which stereopsis becomes possible is in the visual cortex. In V1, although some cells are monocularly-driven, others are binocularly-driven – they have a receptive field in both the left and right eyes. In these cases, the two receptive fields are always of the same type (e.g. selective for the same orientation) and always *roughly* in equivalent positions in the two eyes. In addition to cortical columns being selective for orientation (Section 4.3.1), columns are also selective for **ocular dominance** – the extent to which the cells in the column are driven by inputs from one of the eyes or from both of them (see Figure 14 of Chapter 12 of the Reader, *From retina to cortex*).

Figure 4.12 shows how the images from the two eyes can be used to recover information about relative distance. The top of each panel shows two objects, viewed from above, and their projection into the left and right images. The bottom of each panel shows how the left and right images might actually appear. Note first that the images are reversed; in the world, the round object is on the left but, in the images, it is on the right. Now compare the distance between the round and square images in the left and right eyes. Note that in (a), where the two objects are at the same distance, the distance between the images is the same in both eyes. In (b), where the square object is closer, the image distance is greater in the left eye than in the right, and in (c), where the square object is further, the image distance is smaller in the left eye than the right. The difference in relative positions of images in the two eyes is called **binocular disparity**, and Figure 4.12 demonstrates that it can be used to recover relative distance because it varies smoothly and systematically as the relative distance of the round and square objects is changed. (This is the principle behind the stereoscopic model of rhodopsin in Figure 3.24.)

Disparity can, in principle, be measured by binocularly-driven cells in the visual cortex. For each position in one eye, say the position of the square image in the right eye in Figure 4.12, there are many binocularly-driven cells with a receptive field in that position. And each of these cells has a receptive field at a slightly different position in the other eye, in this case the left. The cell with its receptive field at exactly the equivalent positions in the left and right eyes is, in effect, a 'zero-disparity' cell. It will respond most when the two square images fall exactly in register to its two receptive fields, and this will happen when the square object is at the same distance as the round object (Figure 4.12a). Cells with receptive fields at slightly different positions in the left eye can signal different directions and amounts of disparity. So, for example, a cell with its left eye receptive field at the same position as the square image in Figure 4.12b can signal that the square object is closer than the round object. And a cell with its left eye receptive field at the same position as the square image in Figure 4.12c can signal that the square object is further than the round object.

Motion parallax and binocular disparity can both be measured directly from images and both provide information about relative distance. You can easily demonstrate to yourself how important these two depth cues are to our conscious experience of the world. If you close one eye and keep your head quite still for a few moments, you

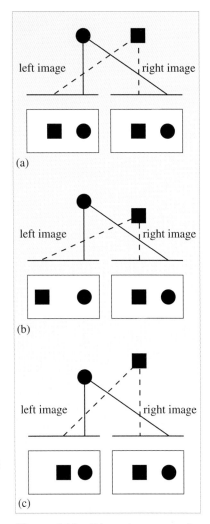

Figure 4.12 Binocular stereopsis. The top part of each panel shows the 3D stimulus arrangement viewed from above. The bottom part of each panel shows the resulting left and right eye images. In the top part of each panel, the point where the rays from the two objects cross for each eye remains constant, corresponding to the lens. Note that the relative positions of the two images in each eye depend upon the relative distance of the objects. These differences in relative retinal positions are called binocular disparity.

deprive yourself of both cues. Under these circumstances, you must rely primarily upon other depth cues and you can still infer the relative distances of objects. If you now move your head from side to side you can restore motion parallax, or by opening both eyes, you can restore binocular stereopsis. For most of us, either of these cues is by itself enough actually to *experience* depth, although perhaps as many as 5% of people do not have full binocular stereopsis, sometimes without knowing it, and they must presumably rely primarily upon motion parallax for their subjective experience of depth.

The stereograms described in Block 1 rely upon binocular stereopsis and provide a good example of how this cue alone can provide a compelling subjective experience of depth. If you are one of the many people who cannot experience depth with SIRDS, this does not necessarily mean that you have anomalous stereoscopic vision. These particular stimuli require inappropriate convergence movements of the eyes, which many of us, not surprisingly, are unable or unwilling to learn.

4.4 Putting it all together

In Section 4.3, we considered how different sub-modalities of the visual stimulus are at first processed by different specialized mechanisms. In this section we change the emphasis to consider how the results of these separate processes are combined into a more useful, and more structured description of the visual stimulus.

4.4.1 Grouping and segmentation

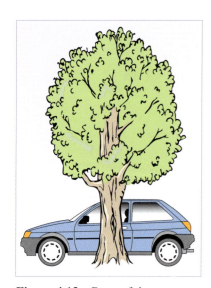

Figure 4.13 Parts of the same object do not necessarily appear next to each other in the image.

It should be clear from Section 4.3 that the initial stages of visual processing are very localized – cells in the retina and visual cortex have small receptive fields and so each can only signal information about one small region of the image. The initial code, then, is very like a set of jigsaw pieces, each piece providing information about the spatial structure, contrast, colour, motion and relative distance of one tiny fragment of the overall picture. Even though the visual cortex is retinotopically mapped, so that the pieces are already in roughly the right positions, the neural jigsaw still needs to be solved. For example, as demonstrated in Figure 4.13, different parts of the image do not necessarily belong together just because they happen to be next to each other. The visual system therefore needs processes for imposing structure upon its initial, fragmentary representation by deciding which bits of the neural jigsaw fit together (**grouping**) and which do not (**segmentation**). You can demonstrate these processes to yourself by looking at a detuned television, which provides a convenient, completely unstructured display of random dots. Although there is no physical structure present, after a few moments your visual system will automatically impose one; most people report seeing rotating swarms of 'flies' against a random background.

As described above, grouping and segmentation processes solve the problem of which bits of the image belong together and which do not. In addition, the resulting groups may have new properties that are not explicit in the ungrouped input. In Figure 4.14, for example, the individual dots have locations, whereas the groups of dots have the additional property of orientation, and the group of these groups has the additional property of shape. Descriptions like orientation and shape clearly make subsequent tasks, like object recognition, easier. Texture is one particularly important example of an additional property that is not present in the original fragments but which is revealed by grouping. Knowing about texture can help in

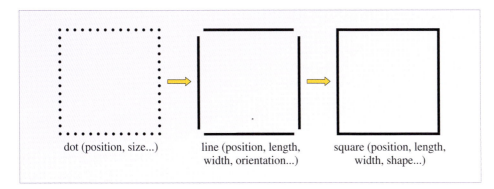

Figure 4.14 The logic of visual grouping. When individual features of the image are grouped together they reveal new descriptive dimensions. For example, individual elements have position, but a group of such elements also has orientation.

object recognition (since a particular texture may be characteristic of a particular object); it can reveal new object *boundaries* that would not show up as simple luminance edges (Figure 4.15a); and it may reveal texture *gradients*, which are obviously important in recovering information about depth, especially the three-dimensional layout of surfaces in the world (Figure 4.15b).

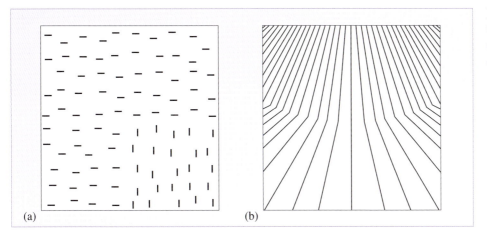

Figure 4.15 The usefulness of texture. Texture can reveal edges (a) and provide information about distance (b).

During the 1920s and 1930s, a group of psychologists in Germany developed a new holistic approach to perception that became known as **Gestalt psychology**. Their research provided a set of rules – the **laws of Prägnanz** – that identify the main factors determining whether or not basic image elements will be grouped together. Some of the most important rules are illustrated in Figure 4.16 (overleaf), although this is inevitably confined to static factors. In practice, one of the most powerful factors is **common fate**, which is the idea that elements moving in the same direction (or towards a common goal) should be grouped together. It should be clear from Figure 4.16 (overleaf) that many of the grouping rules are well matched to the kind of information made available by earlier processes; similarity, for example, may depend upon the size, orientation, contrast, colour, or direction of motion of individual image elements. As you would expect, the processes that solve the neural jigsaw are well matched to the properties of the jigsaw pieces.

Although the neurophysiological basis of grouping is not yet fully understood, it must presumably involve interactions between cortical units with receptive fields in different positions, i.e. between hypercolumns. Such interactions have indeed been found in the cortex and there is evidence that a cell in one hypercolumn may be either excited or inhibited by the response of cells in other hypercolumns. Perhaps

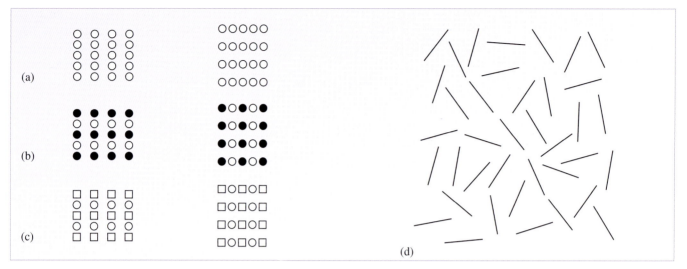

Figure 4.16   Typical Gestalt grouping rules. Grouping by proximity (a), similarity (b, c), and good continuation (d).

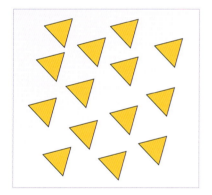

Figure 4.17   These triangles can appear to point in any of three directions, but all of them always point in the same direction.

excitation is needed for grouping – so that cells in different positions with similar properties may cooperate with each other, whereas inhibition may be needed for segmentation – so that cells in different positions with different properties may compete with each other. Cooperative and competitive spatial interactions between orientation-selective cells might, for example, explain the fact that the triangles in Figure 4.17 can be seen to point in any of three different directions; but they all seem to point the same way (cooperation), and they never point in more than one way at the same time (competition). Similar cooperative and competitive interactions between orientation-selective cells might respectively explain the grouping and the segmentation apparent in Figure 4.15a.

4.4.2 Making sense of the initial code

Figure 4.18a shows a very simple scene, which you should have no difficulty in seeing as a dark, roof-shaped wedge resting on a lighter floor plane. How might the processes we have studied so far be involved in making sense of such a simple scene?

The retinal and cortical processes described in Section 4.3.1 should certainly find the luminance edges corresponding to the luminance changes, and might derive a neural line drawing like that shown in Figure 4.18b. This line drawing needs to be grouped into the lines associated with the edges of the wedge and segmented into those associated with the shadow it casts on the floor. If we had additional information about colour, this would help if, as seems likely, the wedge and the floor were of different colours. And in viewing a *real* scene, simple measures of depth information from stereopsis, motion parallax, accommodation and convergence would help by providing information about the three-dimensional layout of the surfaces. But the fact remains that Figure 4.18a is a flat, grayscale image and that, even without colour and depth information, we can still make sense of it. In this case, grouping would presumably be largely on the basis of similarity, described in Section 4.4.1. Lines that are similar in orientation would be grouped together and, since the resulting sub-groups of parallel lines are all joined together, this would allow us to pull together most of the features of the wedge, leaving only the small triangle at the bottom of the figure, which forms the shadow of the wedge.

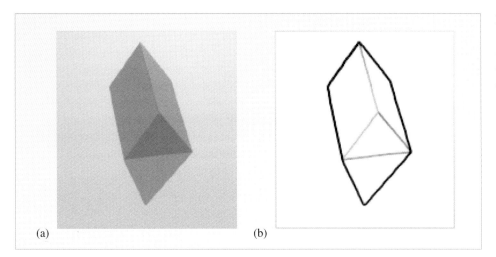

Figure 4.18   The image of a simple scene. (a) Original black and white image. (b) Lines corresponding to the luminance edges in (a).

Now consider the luminance information available in Figure 4.18a. According to the processes described Section 4.3.2, luminance edges provide useful information about changes in surface reflectance. This is true for some of the luminance edges in Figure 4.18a, for example the edges between the far end of the wedge and the floor are due to reflectance changes. But we do not see the *internal* edges of the wedge as changes in reflectance. Rather, we see the wedge as being of constant reflectance, and correctly interpret the changes as different illuminations. In general, the simple processes described in Section 4.3.2 rely on the assumption that illumination is constant, and that reflectance changes across thc edge. As Figure 4.18a reveals, in real three-dimensional scenes, often the reverse is true; reflectance remains constant while illumination changes across the edge.

These changes in illumination, which frequently occur in natural images, are of course due to the three-dimensional layout of the surfaces. A surface that is at right angles to the direction of the light source will be brightly illuminated because it can 'catch' all the light from the source, whereas a surface that is exactly parallel to the direction of the source will not be illuminated at all, because it cannot 'catch' any of the light flooding past it. As the angle between the surface and the light source varies, the amount of illumination will vary smoothly between these two extremes. Such differences in angle account for the different illumination of the three sides of the wedge in Figure 4.18a.

○   Try to identify another factor causing a difference in illumination in Figure 4.18a.

●   The shadow cast on the floor is poorly illuminated because the wedge prevents the light from reaching it. In the real world, objects often shade each other like this and, sometimes, one part of an object shades another part of itself.

On the one hand, changes in illumination are potentially useful to the visual system because they can provide important information about surface layout in the world; that is why shading is such an important depth cue (Section 3 of Block 1). On the other hand, however, they raise an important complication, for how does the visual system know whether to interpret luminance changes in the image as changes in reflectance, changes in surface angle or shadows?

In practice, the scale of the luminance change in the image can help make this distinction because changes in reflectance often produce sharp luminance edges, changes in surface angle often produce smoother and broader luminance edges, and shadow edges tend to be intermediate in scale. And as we know from Chapter 14 of the Reader, *Spatial vision*, the visual system is well equipped to measure the scale of luminance edges. But, these general guidelines are not *always* true. In Figure 4.18a, for example, almost all of the luminance edges in the image are sharp, though almost all of them are due to changes in surface orientation. Manufactured objects like wedges tend to have sharper corners and produce sharper luminance edges than do natural objects like trees and people.

In general, to make sense of even simple scenes, it seems likely that the visual system must combine information from different sub-modalities, exploring different possible interpretations in an attempt to find the most coherent explanation of the image. In Figure 4.18a, for example, grouping processes based on the spatial structure of the image might suggest that most of the edges belong to the same object, and we might start by assuming that objects tend to have constant reflectance. Suppose we then assume that the remaining edges in the image (the small triangle at the bottom of Figure 4.18b) are part of a shadow; after all, the triangle has roughly the same shape as one of the possible object regions and is attached to it along one edge. On this assumption, the relationship between the apex of the shadow and the part of the object that produces that apex would suggest that the light source is to the right and slightly backward in the scene. And a light source from that direction could explain all the luminance changes in the image as three surfaces arranged as a wedge. This series of assumptions, each suggested by some aspect of the image, provides a simple and coherent explanation of the image; one that is more plausible, for example, than four strangely shaped bits of differently reflective card laid flat on the floor, even though this unlikely arrangement would produce exactly the same image.

Exactly the same kind of process presumably underpins the illusion of bumps and hollows shown in Figures 3.2 and 3.3 of Block 1. Here we interpret all the luminance edges in the image as being due to changes in surface orientation. In the absence of any other information about the direction of the light source, we make the reasonable assumption that light comes from above – and that assumption leads us to interpret the circular features of the figures as either sticking out of or into the scene.

STUDY FILE

Activity 4.2 Visual illusions

The CD-ROM provides illustrations of some other visual illusions. Many of these illusions seem to provide evidence of the visual system exploring possible assumptions in trying to make sense of an image. Such illusions are useful in illustrating some aspect of visual processing, but we should never lose sight of the fact that even though illusions are mistakes, they illustrate processes that are generally sensible and necessary when trying to make sense of the natural visual world. Further instructions are given in the Block 4 *Study File*.

4.5 The next step

In Section 4, we have dealt exclusively with the initial, descriptive aspects of visual perception and with what are often called bottom-up processes. We have also been guided almost exclusively by one particular visual task, that of visual object recognition. Section 5 continues this theme by describing the subsequent, inferential aspects of vision, often called top-down processes, again emphasizing the task of visual object recognition and, in particular, the specific examples of reading and of face recognition.

Although it is convenient to deal with bottom-up and top-down processes in different sections, it is crucial to remember that neither can work without the other. This is obviously true for the visual system, but it is no less true for the vision scientist and for the student trying to understand how the visual system works. As we shall see in Section 5, an understanding of low-level descriptive processes can help us to formulate plausible theories about subsequent inferential processes. And, as is hopefully clear from your reading of Section 4, the understanding of low-level processes begins only with an understanding of the high-level task that the visual system is trying to perform. The task defines the problems and, without an understanding of the problems, we cannot hope to understand the solutions that the visual system has evolved.

Question 4.1

What might be the role of an individual simple cell in encoding the spatial structure of the image?

Question 4.2

What is likely to happen to our perception of the lightness and brightness of a scene when the sun disappears behind a cloud?

Question 4.3

What might have been the visual motivation for changing the traditional UK colouring of flexible electrical cables (red = live, black = neutral, green = earth) to the present convention (brown = live, blue = neutral, green/yellow = earth)?

Question 4.4

Which is the better cue to relative distance: motion parallax or binocular stereopsis?

Question 4.5

Motion provides an important grouping cue. Is the Gestalt rule of 'common fate' (the notion that things moving towards a common goal should be grouped together) an adequate basis for such grouping?

Question 4.6

What kinds of bottom-up information might be useful for recognizing a chair?

4.6 Summary of Section 4

We started this section, as does the visual system, by considering different sub-modalities of vision. By considering the information potentially available from, for example, the spatial structure or pattern of wavelengths available in the image, we gain insight into the sorts of problem that the visual system is trying to solve and thus into the biological processes that are found in the retina and cortex.

Many of these processes seem to be well adapted to solving one particular problem, to recovering surface reflectance despite changes in illumination, for example. And many of them seem consequently to be based on some general assumption about the world, that illumination remains constant across edges whereas reflectance does not, for example.

We cannot conclude, however, that these early processes, good as they are, actually *solve* the problems of vision. It is clear that even simple scenes like the one depicted in Figure 4.18a require information to be combined across different positions in the image and across different visual sub-modalities. Knowing about surface colour, for example, can help us to decide which bits of the scene belong with which, whilst knowing about the three-dimensional surface layout can help us decide if the general assumption about reflectance changes is correct.

So the role of retinal and cortical processing is not to solve visual problems but to provide information that makes the subsequent solution easier by picking out the aspects of the image that are likely to be useful, and by describing them in a way that makes them easier to interpret.

5

Recognizing shapes

So far we have considered how lines and edges, boundaries and surfaces are detected, but those are not the components of a visual scene that we normally notice. We don't see lines: we see print; we don't see surfaces: we see faces. In this section we shall look at the processes involved at this more complex level of perception, and will use face recognition and reading as examples. These are particularly illuminating topics, as the first appears to be 'natural' while the second is learned, so they cover the spectrum of our abilities. In other words, as social animals, it is reasonable to assume that we evolved mechanisms to allow us to recognize other members of our group. In contrast, writing was invented only a few thousand years ago, so we cannot have evolved brain structures specialized in reading.

5.1 The beginnings of a theory

Letters and words are good examples of relatively complicated shapes that lend themselves to theorizing. Take the letter 'A' as an example; one can easily imagine how different line detecting neurons would respond to the three straight lines that make up the letter. Other neurons would take their inputs from the outputs of the line detectors, and if the appropriate three inputs were active, then one of the neurons would indicate that there was an 'A' in the visual scene. An idea like this was first proposed about half a century ago. It was called the **pandemonium model** and is represented in Figure 5.1, which demonstrates how a letter might be detected. As well as meaning an awful lot of noise, pandemonium literally means 'all devils'. Here, each little devil is a simple creature (equivalent to a neuron, which by itself cannot be clever) which is capable of only two things: listening out for particular simple signals (lines, angles, etc.) and shouting. The demon shouts louder if it is more certain that its own special signals are present. (All this shouting obviously leads to pandemonium!) At the top of the decision tree the information should have been narrowed down to one possibility, and so the input has been analysed.

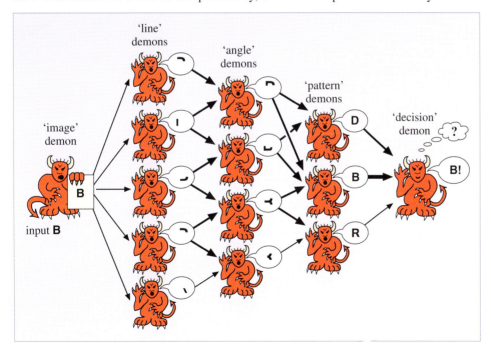

Figure 5.1 The Pandemonium model of perception. A hierarchy of simple 'demons' makes increasingly complex decisions about an input pattern.

This ingenious, and certainly engaging model has a number of problems, as we shall see, but it is not completely at variance with the most recent ideas.

5.2 Grandmother cells

There are fundamental problems with any model that implies the existence of a hierarchy of cells, narrowing down to a single cell that encapsulates the results of the recognition process. The concept has been summed up by postulating the existence of a 'grandmother cell'. This would be a neuron in your head, which became active when your eyes received an image of a nice grey-haired old lady, wearing a shawl and sitting by the fire in her rocking chair (if that is your image of a Granny). The first danger of such a system lies in the mortality of neurons. They do die from time to time, so what if your grandmother cell died? You would never recognize her again! As will be described later, there have been cases where people have lost very large numbers of brain cells, through stroke or other injury, resulting in an inability to recognize a close relation. However, there is absolutely no evidence that the loss of a single cell causes catastrophic impairment of that sort.

The other difficulty with having Granny represented by a single neuron is the question of what good the neuron would be to you. Somehow the rest of the brain would have to 'see' the neuron, so that it could, for example, trigger memories of her telling you stories when you were small, or of her delicious apple pie. If the 'apple pie' and 'story' neurons always became active when the grandmother cell was active, there would no longer be a unique grandmother cell. There would be a whole group of neurons associated with spotting Granny.

5.3 The timing problem

How long does it take to recognize a word or letter? There is a technique that has been used in all sorts of psychological research, called the **lexical decision task**. The adjective 'lexical' means to do with words, and the task consists simply of deciding whether some letters spell a word or not; a computer is often used. The computer flashes up some letters on the screen, and as fast as possible the participant has to decide whether it is a word, and press one of two keys, to indicate 'yes' or 'no'. The computer times how long it takes from the letters appearing to the key being pressed.

Suppose you saw BRANE: you would press 'no' (one hopes). To do that, your eye has first to get the cones firing, and their information has to be relayed up to all the line detectors. Suppose the information then had to go through a hierarchy of neurons, travelling across synapse after synapse, as a pandemonium-like analysis took place. That would need to happen for five letters, and having decided what they were, there would have to be other 'demons' that knew how to assemble them to make words. You will have a vocabulary of some tens of thousands of words to check, and you will have to be sure that, although these letters sound like BRAIN, this version is not one of the words you know. Having made that decision, you would then need to get a message across to neurons in your motor cortex (the part of the brain that controls movement), so that they could send signals all the way down the long fibres that pass down your spine and out along the arm, to produce a key press.

How long would all that take? The way we have described it, the answer would probably be several seconds; in fact your response would probably be made in somewhere between a half and two-thirds of a second. We know that a typical synapse between neurons takes an appreciable fraction of a second to transmit information, so the only way a decision could be reached this quickly would be if only relatively few synapses were crossed. Clearly, this might barely get us beyond the first level of demons, so it must be the case that the analysis does not take place through such a 'tall' hierarchy of decisions. Instead, a very large amount of processing must take place in parallel (**parallel processing**). In this way even millions of synapses could be involved, but they would work simultaneously, in parallel, rather than each waiting for the last to finish. We will return to this idea, when we consider modern theories of perception.

5.4 Reading for meaning

So far we have considered the kind of neural connections that might be able to analyse a printed word, but there is more to reading than analysing a pattern of lines: we read to understand. Without that further step, looking at print would be like examining an unknown foreign language. Somehow, the initial analysis of the letter patterns moves on, to activate the same mechanisms that understand the spoken word.

The processes involved are described in Chapter 17 of the Reader, *The perception of words* by Peter Naish, which you should read now.

Chapter 17 raises a number of issues, such as the effect of context upon analysis, and whether the brain really does work in parallel. We will return to these later, but first we turn to the process of face recognition, which raises similar issues. We can address these at the end.

5.5 The perception of faces

Whereas reading feels effortless and automatic, we often experience difficulty in recognizing faces. Most of us must have had the embarrassment of meeting someone whose name we clearly should know, but that we can't for the life of us remember!

○ It seems strange that something we are supposed to have evolved to do can be quite difficult, while the acquired process of reading is easy. Why might that be?

● We would not have evolved to learn names, only to recognize different individuals.

Even when we cannot remember a name, we do usually know that the person is familiar. We are astonishingly good at that, and after just one exposure to a face, can often pick it out again as being familiar, even years later. Contrast that with children learning to read; how much easier if they recognized words after one exposure! Reading only becomes easy after considerable practice.

5.6 'Don't I know you from somewhere?'

There is an intermediate stage between recognizing that a face is familiar and naming its owner; this stage is the recollection of biographical detail. Recognition appears always to follow the sequence: familiarity, personal details, name. The end point (naming) is not always achieved, but if it is, it cannot normally be reached without passing through the earlier steps. Thus, someone might well say, 'Oh, that's 'what's-his-name', you know, he acted in *Silence of the Lambs*.' On the other hand, it would be most unlikely to find a person saying, 'That's Anthony Hopkins, but I can't remember what he does for a living.' Our success at finding a name is strongly influenced by context, since that can aid or confuse the processes that find the biographical detail, which, as stated, is a necessary precursor to naming. Thus, if we regularly meet someone in one situation, but one day see them in a different setting, we are quite likely to be unable to remember their name, although we will recognize that they are familiar.

Recognizing that a face is familiar can be considered the heart of face perception. Large areas of the brain are involved in face processing, but brain mapping has shown registration of familiarity to be particularly associated with a region known as the **fusiform gyrus**, which lies ventrally, between the occipital and temporal lobes. Brain damage can lead to a failure of recognition; a sufferer may distinguish faces, but be unable to identify them, or even judge whether they should be familiar. The condition is termed **prosopagnosia**, and it can be quite distressing, as, for example, when a husband suffers a stroke, with the result that he no longer recognizes his wife. This is not a memory loss, simply requiring him to learn who she is once more; there is permanent damage to the ability to experience familiarity. In one curious case of this sort, the sufferer retired (imagine his difficulty in the workplace, if colleagues remained permanent strangers), and took up sheep farming. Although he never regained an ability to recognize people, he did develop one for identifying his sheep, and could tell them apart! In another example, a one-day-old infant suffered brain injury. The brain at this age often appears plastic, so, with one part damaged, another will take on the processing. However, this unfortunate child grew up to be prosopagnosic, although he was able to identify objects other than faces. Cases such as these strongly support the claim that face processing utilizes specially evolved mechanisms within our brains.

5.7 Facial analysis

The visual pattern that makes up a face can be processed in two ways: the individual features can be analysed, in a rather piecemeal manner, or the overall pattern can be used.

○ To what word recognition processes is this distinction analogous?

● It is rather like the alternatives of either translating from letters to sounds, or simply recognizing the entire word pattern.

Just as there is evidence that skilled readers tend to process words by shape, so children seem to move from feature processing to a more holistic strategy, as they grow older. We can all process at the feature level, and indeed can sometimes recognize a famous face from just the eyes or the mouth. However, it is probably quicker to use all the detail simultaneously (notice, parallel processing again), including information in the face layout, such as the space between the nose and the mouth. As we appear to need to learn this whole-face approach, it is likely that it is not fully 'pre-programmed' by evolution. Evidence for this comes from the way we respond to inverted faces. Almost all our learning takes place with upright faces, so when confronted with an inverted image we revert to processing by feature, without registering the relationships *between* the features. The results can be strange, as you will see by examining Figure 5.2.

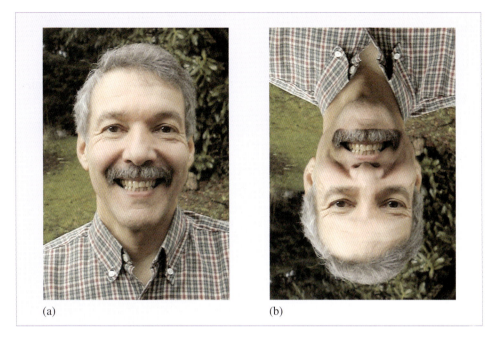

Figure 5.2 (a) An upright picture of Peter Naish (one of the SD329 authors), and (b) an apparently inverted version of the same picture. But try turning the page upside down.

(a)   (b)

The upside down face in Figure 5.2b has had two sections inverted; one around the eyes, and the other a rectangle containing the mouth. As you examine the picture, you will be carrying out what is known as **mental rotation**, to gain an impression of an upright face. You do this, feature by feature, *as required*. Rotation is required for his nose, and also for the face outline, to put his hair at the top and chin at the bottom. It is not required for the eyes or mouth, but, since rotation is not a conscious process, the viewer is unaware of the differences in processing. The end result is simply a set of features, which look very much like those of Figure 5.2a. When the basic face shape of Figure 5.2b is placed upright, the grotesque layout of the features becomes apparent. This unflattering manipulation was first reported for a picture of Margaret Thatcher, so the effect has become known as the **Thatcher illusion**.

5.8 Processing in context

In both word and face perception we have met examples of context effects. Of course, these effects are apparent in all areas of perception; Figure 5.3 (overleaf) shows an example of **size constancy**, which is a context-driven effect. The converging lines of the hedges and road signal distance (see Block 1, Section 3), and this context results in objects further away being perceived as larger than might be expected, given their actual image sizes on the retina. Our brains produce perceptions that are closer to 'how the world is', rather than faithful transcriptions of what the sense organs register.

Once we have learned to interpret converging lines as distance cues, the size constancy process becomes automatic. This results in effects such as the **Ponzo illusion**, shown in Figure 5.4 (overleaf).

○ Why do people seen looming out of the fog sometimes look taller than life size?

● Haziness is usually (when there is no fog) an indication of distance, so it triggers the size constancy process. If the people are actually quite near, their images will already be reasonably large – the 'size enhancement' results in them looking too tall.

Figure 5.3 The effect of size constancy. The tiny feature in the road, at the tip of the arrow marking, is a replica of the on-coming car in the distance.

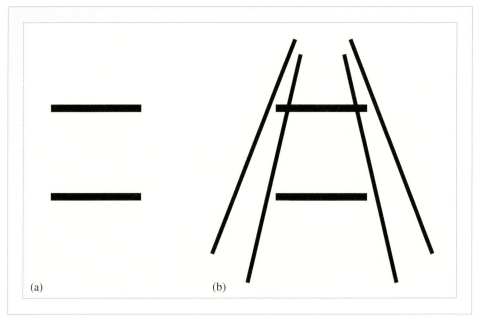

Figure 5.4 The Ponzo illusion. The upper and lower horizontal lines are of equal length in both (a) and (b), but in (b) the upper looks slightly longer. The surrounding lines induce an impression of perspective, suggesting that the upper line is more distant.

5.9 Bottom-up and top-down processing

Although *mis*perceptions can sometimes result from context effects, it is clearly an advantage that additional information can influence processing. As we have seen, it can result in faster processing and better representations of the environment. What is not immediately clear, is how context can achieve these results. The kind of processing suggested by a model such as Pandemonium leaves no room for any influence from other aspects of the scene, or from previously learned material. It simply works on the incoming information, which is passed up a decision tree, until the analysis is complete; the outcome is determined entirely by the signals to be processed. Because of these characteristics, this kind of analysis is referred to as bottom-up, or **data driven**. The earliest stages of processing, from ganglion cells to primary visual cortex, do have this quality, but it is clear that there is eventually an influence from the other direction.

Context effects imply that what is already known, either from previous experience, or from other aspects of the situation, exerts an influence upon how a stimulus is finally interpreted. This process is described as top-down, or **concept driven**. One of the earliest theorists to popularize the top-down approach was British psychologist Richard Gregory. Throughout his career he has had an interest in visual illusions, because the kinds of errors a system makes can be very revealing about how it works when things are going right. Gregory suggested that visual perception is effectively a process of hypothesis testing. A scientist tests a hypothesis when first developing an idea (a hypothesis), making predictions that would follow if the hypothesis were correct. The scientist gathers data to test the predictions; if they prove correct, then the hypothesis is supported (although not necessarily *proved*), but if the predictions turn out to be wrong, then a new hypothesis has to be developed. Gregory conceived of the brain as gathering data (via the eyes, say), and beginning to develop a hypothesis as to what might be 'out there' to explain the particular pattern of information. On the basis of the hypothesis, more specific information would be sought, in an attempt to confirm it. If essential information were missing, or did not correspond with the expectation, then the hypothesis would be abandoned and a new interpretation attempted.

Notice that this explanation of perception does not say that every last detail has to be tested; once the data seemed to be fitting well, the brain could produce a conscious perception, without taking time to check any further. This might explain why we sometimes see what we expect to see, rather than what is really there. Some visual stimuli can be ambiguous; the **Necker cube**, shown in Figure 5.5 is an example. The corner with a dot beside it can be interpreted as being the nearest corner to the observer, or the furthest. If you keep staring at the figure you will find that the two interpretations alternate. It is as if the brain finds one hypothesis, but, because you keep looking, it then finds another. So called 'impossible figures', such as the one illustrated in Figure 5.6, demand two conflicting hypotheses simultaneously, so are very disturbing! This example requires you to see the top of the three-dimensional Ψ sign as having three prongs, yet lower down there seem only to be two.

These two examples do give the impression of alternately forming and abandoning hypotheses. However, this may be unusual, since, at least at a conscious level, more often than not we seem to form the correct perception right away. This, as with so many other aspects of perception, could be accounted for by parallel processing; different hypotheses would be tested and developed simultaneously. A complete account of perception must incorporate the ideas of parallel processing, together with top-down mechanisms.

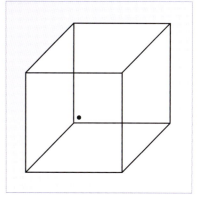

Figure 5.5   The Necker cube. The corner with a spot can seem the nearest or the furthest point of the cube. With persistent viewing it will switch between the two.

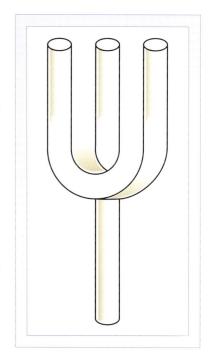

Figure 5.6   An impossible figure. The prongs seem to number two or three.

5.10 The connectionist account

Recently the field of **cognitive psychology** has been considerably influenced by a methodology that attempts to use computers to mimic the rudiments of neural architecture. As this is a very broad course, no attempt will be made to examine the ideas in depth; instead, a brief outline of the concepts will be given. They are based on two observations: first that a great deal of neural processing takes place in parallel, and second that any one neuron in the brain may have many thousands of synapses with others. These interconnections form networks of neurons, which are sometimes referred to as '**neural nets**'. Because there are so many synapses in a net, there is an enormous number of possible combinations of neural firing, and the same set of neurons can end up with many different patterns of activity. The different patterns could represent different 'states'. Presumably one such pattern would represent the grandmother concept; not a grandmother cell, but a grandmother net. Since Granny is represented in this way, she will not be 'lost' if one neuron dies; the information is said to be 'distributed'. Because the neurons in the net all do their processing in parallel, the overall mechanism is referred to as '**parallel distributed processing**', often abbreviated to **PDP**.

The synapses are the driving force of these nets. Whether or not one neuron has much influence on another to which it is linked depends upon the *strength* of the link. In other words it depends upon how effective the synapse is at getting information across. This effectiveness can be modified, and when a sending and a receiving neuron are both active there is a tendency for the synapse to be strengthened, so that the next time the first neuron fires it will be more likely to trigger the other. This is the essence of learning in the net; repeated activity strengthens the synapses involved. After learning, some synapses will be strong and others weak. This pattern of synaptic strengths holds the key to how the net will respond to different incoming signals. Because it is the connections that are all important in these nets, the theories based on this idea are described as '**connectionist**'.

Figure 5.7 is a highly simplified schematic diagram, to give an impression of what has been described. It is not intended to imply that this is how letters are actually recognized. It shows seven axons (the long horizontal lines) which have come from simple feature-detecting neurons (which are not shown). Only a few of the possible features are represented, and they are placed as labels beside the corresponding axons. For example, the third axon down comes from a horizontal line detector, and it will be active if a horizontal line is in the field of view. Each of the incoming axons has contact with the dendrites of letter-detecting neurons; just four are shown. Where a strong synapse has formed, this is indicated with a dot at the crossing point. If sufficient incoming activity is passed via the synapses to a letter-detecting neuron it will fire. Let us suppose that it takes three active synapses to make the neuron trigger. If the second, third, sixth and seventh feature axons were active the four synapses involved would trigger the 'A' neuron. Even if one of the incoming signals were absent (the little angled junction perhaps) there would still be sufficient activity to cause the 'A' to be signalled. One could describe this as the effect of context, since the other components that make up an 'A' are present. With so many connections in an actual brain, it would clearly be possible to include influence from the perception of other letters, or whole words. Similarly, the networks that respond to faces could be influenced by others responding to environmental cues.

There are several observations that support the connectionist account. First, from microscopic neuro-anatomical investigation, it is known that the brain contains

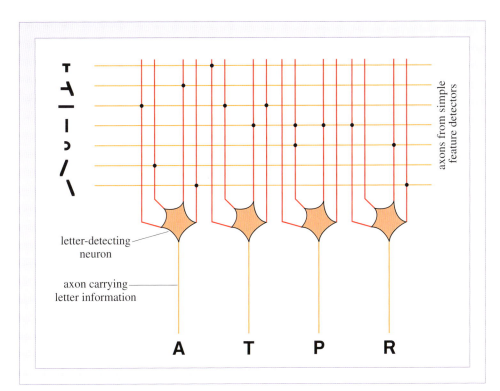

Figure 5.7 A stylized neural circuit for recognizing letters from their features.

letter-detecting neuron

axon carrying letter information

axons from simple feature detectors

A   T   P   R

regions where neurons are massively interconnected (Figure 5.8). Moreover, brain-mapping studies show that activity arising from a partial stimulus can lead to activity in another part of the brain, a part that would normally be involved if the stimulus were complete. (More details of this effect are given in Block 7.) As mentioned above, connectionism has been tested by computer modelling. It is not possible to mimic the vast numbers of neurons and synapses to be found in a real brain, but researchers have set up computers to simulate a few hundred interconnected cells. Early results are very encouraging, with networks able to perform rudimentary face and word recognition tasks. Like humans, the nets have to learn and, significantly, the kinds of errors they make while doing so are often similar to those produced by children. Humans can suffer brain injury, sometimes resulting in **dyslexia** or prosopagnosia; simulated nets can also be damaged, and the resultant errors are also like those of human patients.

It should be noted that, in its basic form, the simulated neural net begins as an amorphous set of interconnected 'neurons'; all that they achieve has to be attained by learning. While that is generally also the case for the human brain, there is clearly a degree of 'pre-wiring' – we have evolved neural circuits that are predisposed to process certain stimuli, and different processes take place in different regions.

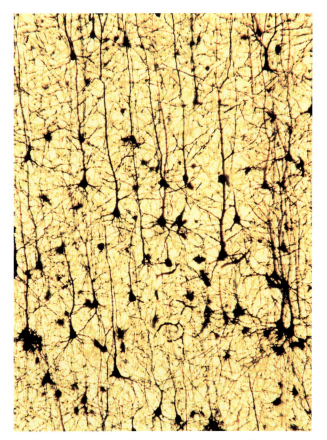

Figure 5.8 Neurons in the cortex, showing the density of interconnections.

○   What example of pre-wiring have you met in the preceding account?

●   The ability to perceive faces appears to be predetermined, at least to some extent.

It has also been claimed that some phobias do not have to be learned. For example, many people have a fear of spiders, for no *logical* reason; it is possible that, early in our evolution, the development of a fear of small scurrying creatures had survival value. It is interesting that, in Britain, one is far more likely to die by being run over by a bus, than through being bitten by a spider. Nevertheless, bus phobia is almost non-existent, because, of course, there were no buses around during our evolution!

Question 5.1

Faces from other races can sometimes look very similar to one another, while those of our own background appear distinctive. What might this observation tell us about how well evolution has prepared us to recognize faces?

Question 5.2

How might a skilled reader be expected to process words that have been printed upside down?

Question 5.3

Consider the printed phrase: 'His poor eye sight'. What factors will be involved in the processing of these printed words, and which words might be recognized the fastest?

5.11 Summary of Section 5

Visual perception is almost entirely an automatic process, but that does not mean that the conscious outcome (perceiving a face for example) is determined entirely by the nature of the stimulus. There is a wealth of evidence that other factors, such as context or prior knowledge, can influence the way in which a stimulus is analysed and recognized. The ability of complex interactions to determine the final perception can be accounted for by the extensive interconnections in the brain; these are the same interconnections that make parallel processing possible, and without which our recognition processes would be impossibly slow. Much of seeing is based upon learning, although we may be predisposed to recognize certain types of stimuli, and hence be quick at identifying them.

Perhaps the most important aspect of perception (not only visual) is that it is a pragmatic process; at the conscious level it is the brain's 'best guess' as to what the world is like. We do not process the stimulus in order to reproduce the pattern of activity in the sense organ: we process for understanding and recognition.

Objectives for Block 4

Now that you have completed this block and Chapters 8 to 17 of the Reader, you should be able to:

1   Define and use, or recognize definitions and applications of, each of the terms printed in **bold** in the text.

2   Demonstrate an understanding of the properties of light, including the relationship between wavelength and colour, refraction, diffraction, and reflection from surfaces. (*Questions 2.2, 2.4 and 3.2*)

3   (a) Explain additive and subtractive mixing and understand the meaning of the terms hue, value and chroma. (b) Outline the concepts behind the CIE colour diagram. (*Questions 2.1, 2.5 and 2.6*)

4   Explain the concept of spatial frequency and its relevance to imaging systems. (*Question 2.7*)

5   Describe the structure of the human eye.

6   Describe the processes that lead to the formation of an image on the retina and explain the factors affecting the quality of that image. (*Questions 3.2 and 3.4*)

7   Describe the special properties of the cornea. (*Questions 3.1 and 3.3*)

8   Describe the structure of the retina and explain the processes occurring in the retina which convert a light signal into a neural signal. (*Question 3.7*)

9   Demonstrate an understanding of the role played by rods and cones in vision. (*Questions 3.6 and 3.9*)

10 Explain the mechanisms of colour vision and describe the different forms of colour deficiency. (*Questions 3.5, 3.6, 3.8 and 4.3*)

11 Distinguish the different types of movement of the eye and understand their purpose. (*Question 3.10*)

12 Explain the different ways in which the eye can adapt to a wide range of luminance levels. (*Question 3.11*)

13 Describe the different measures of visual acuity and discuss the factors affecting acuity. (*Questions 3.12–3.15*)

14 Explain how the contrast sensitivity function measures visual performance. (*Question 3.16*)

15 Describe the structure of the neural pathway from retina to visual cortex.

16 Explain how retinal and cortical receptive fields are involved in the detection of edges in the image. (*Question 4.1*)

17 Explain the distinction between lightness and brightness and the neural processes that underpin these perceptual dimensions. (*Questions 4.2 and 4.3*)

18 Explain how information about relative distance may be recovered from binocular stereopsis and motion parallax. (*Question 4.4*)

19 Define the significance of visual grouping processes. (*Questions 4.5, 5.1–5.3*)

20 Explain the distinction between bottom-up and top-down processes in vision. (*Questions 4.6, 5.1–5.3*)

Answers to questions

Question 2.1

Value is also known as brightness; chroma can be described as colourfulness or saturation.

Question 2.2

(a) The short wavelength end of the range corresponds to violet.

(b) Red light that is just visible has the longest wavelength, approximately 780 nm. Frequency is given by $f = c/\lambda$ so the frequency is $(3 \times 10^8 \text{ m s}^{-1})/(780 \times 10^{-9} \text{ m}) = 3.8 \times 10^{14}$ Hz.

(c) This is smaller than the frequency of violet light – a longer wavelength corresponds to a lower frequency.

Question 2.3

The wavelength range quoted goes from red down to infrared. The radiation produced by the Sun peaks in our visible range (380–780 nm) and is less at longer wavelengths. So we would 'see' less of the energy produced by the Sun. However, bodies at lower temperatures than the Sun produce more infrared radiation. So we would 'see' other radiant sources, such as electric heaters and cookers, as very bright objects. (We would also see other people, even at night, as the black body radiation produced by a body at the normal body temperature (37 °C) peaks at wavelengths just less than 10 000 nm.)

So how would this affect our lives? Well the illumination that we would need to move around, read, etc., would be of a lower frequency. We would also be able to move around much more easily at night so we might well find that there would be less distinction between night and day. One can speculate what effect this would have on our lives.

As far as vision is concerned there would also be some other more subtle effects to do with diffraction and resolution. You may rightly feel that you will be in a better position to consider these after studying the rest of the block.

Question 2.4

(a) The reflected light is mainly at the blue end of the spectrum so in sunlight the object will appear blue.

(b) There is very little reflectance at 589 nm (the wavelength of sodium streetlights) so the product of the incident light spectrum and the reflectance will be close to zero. The object will appear almost black.

Question 2.5

Mixing red and green light gives yellow light. Red and green are two of the primary colours of light.

Mixing red and green paint, as any playgroup leader will tell you, gives a rather messy brown colour. This is because the red paint absorbs all colours except red; the green paint absorbs all colours except green. Theoretically the mixture of the two colours in paint should give black but, since the original colours are unlikely to be pure, the end result is usually a rather sludgy brown colour.

Question 2.6

(a) False. The colours that lie along any line on the CIE diagram correspond to mixtures of the two colours at the end points of the line. The lower line joins red to violet so the colours are mixtures of red and violet. They are not colours found in the spectrum (these are found around the curved sides of the diagram) so they are known as 'non-spectral purples'.

(b) True. Hue changes around the curved line; chroma (also known as colourfulness or saturation) increases outwards. The value (or brightness) can only be shown in a 3D diagram.

(c) False. *Additive* mixtures of two spectral colours lie on a straight line joining those colours.

Question 2.7

There are many examples you could have chosen. A narrow-striped shirt is an example of a one-dimensional pattern containing high spatial frequencies. A checked shirt contains high spatial frequencies in two perpendicular directions. A shirt with broad stripes (e.g. Newcastle United), viewed from the same distance, gives rise to lower spatial frequencies on the retina (Figure 2.13). Wallpapers can also be used as an illustration. A portcullis gate would give rise to lower spatial frequencies in two dimensions, iron railings in one.

Figure 2.13   Examples of shirts with different spatial frequencies.

Question 3.1

Your list might include:

- smaller radius of curvature than the rest of the eye

- transparent to visible light

- tough

- immune privileged

Question 3.2

Underwater the anterior surface of the cornea is in contact with water rather than air. The refractive index of water is approximately 1.33, so the refractive power of the surface is reduced to less than 3 D, leaving only about 12 D in the remainder of the eye. Wearing goggles restores the power to the normal value; the goggles, of course, have plane surfaces and therefore no focusing effect, but allow the cornea to be in contact with air.

Question 3.3

The centre–centre spacing of the collagen fibrils in the cornea is about 60 nm. This is much less than the wavelength of visible light so destructive interference occurs in all directions except the straight-through direction. This renders the cornea transparent to visible light.

X-rays have very much smaller wavelengths, typically 1 nm or less, so diffraction gives maxima in many different directions, scattering the X-rays. (This phenomenon is actually very useful to scientists because it allows determination of the spacing of the fibres.)

Question 3.4

(a) He suffers from myopia (short sight) as this requires a negative (diverging) corrective lens.

(b) He will need to wear his glasses most when he is looking at distant objects – for example in a lecture theatre, at the cinema, or driving a car.

Question 3.5

Figure 2 in Chapter 11 of the Reader shows that monochromatic light of wavelength 560 nm will elicit no response from the S cone and equal responses from the L and M cones. An equal intensity mixture of the two wavelengths given will also elicit equal responses from the L and M cones. There is therefore no way that the human eye can distinguish the two. Hence the mixture of red and green light is seen as yellow.

Question 3.6

Staring at the blue patch causes the S cones to become fatigued. They will take some time to recover to their normal receptive state. If meanwhile you stare at the yellow patch (or even a piece of white paper) the L and M cones respond as normal but the S cones have a very reduced response. The end effect is that you 'see' the more intense yellow colour that would be seen if the reflected light did not excite the S cones.

Question 3.7

(a) Figures 3b and c of Chapter 10 show that the foveal region is an avascular, rod-free zone.

(b) The interneurons are pushed to the periphery – i.e. they are displaced.

(c) This area is rich in cones. There is little or no convergence in the cone to retinal ganglion cell pathway relaying the neural image to the brain. So this is the area of high visual acuity, where any deflection of light has the potential to distort the image.

Question 3.8

In the auditory system the different parts of the cochlea are each tuned to a different frequency. If the visual system were similarly based on a large number of sensors each tuned to respond to a narrow band of frequencies then we would be able to distinguish between physically different colour signals. Our experience of colour vision and therefore our experience of the world would be very different indeed.

Question 3.9

The visual pigments have maximum absorbance at slightly different wavelengths. Each one consists of retinal bound to an opsin and it is the slightly different structures of the opsins that give rise to the wavelength differences.

Question 3.10

1 Disjunctive movements mean that the eyes both point towards the closer object.

2 The shape of the lens alters so that accommodation occurs.

3 The pupils contract slightly (the near reflex, Section 3.2).

Question 3.11

Rods are well adapted to dark (scotopic) conditions and cones are well adapted to bright (photopic) conditions but neither system functions particularly well under intermediate (mesopic) illumination, such as at dusk.

Question 3.12

From the data given, the angle subtended by the wall at the eye of a spaceman is given by:

$$\tan \theta = \frac{6\ \text{m}}{200\ \text{km}} = \frac{6\ \text{m}}{2 \times 10^5\ \text{m}} = 3 \times 10^{-5}$$

from which $\theta = 1.7 \times 10^{-3}$ degrees = 0.1 minutes.

The visual acuity required is therefore 1/0.1 = 10. This is well above the maximum value achievable by the human eye.

Even taking into account the possibility that the wall might have a shadow on one side, and therefore appear about twice as wide, it would still not be visible. Several astronauts, including Neil Armstrong, have confirmed that this is the case.

It is perhaps also worth commenting on the statement that this is the only possible human artefact visible from space. There are plenty of others that are possible candidates – the Dutch polders are visible and the Egyptian pyramids have larger dimensions than the width of the Great Wall so, theoretically at least, could be seen. Cities too can be seen at night because of all the streetlights.

Question 3.13

Everyone's eyes are different. But most people will find that the oblique test gives a lower acuity than either horizontal or vertical. If you found large differences between horizontal and vertical this suggests that your lens does not have the same refractive properties in the two directions. This is known as astigmatism.

Question 3.14

In the eye, the sensitivity and acuity are far from constant across the whole retina. High sensitivity requires the pooling of responses from many photoreceptors, whereas good acuity depends on access to the responses of individual photoreceptors. The receptive fields of retinal ganglion cells are small in the fovea (giving good acuity) and are larger in the retinal periphery (giving better sensitivity). For example, Figure 3.48 shows that the acuity is considerable reduced away from the fovea. However, the eye compensates for this by rotating the eyes so that, once a stimulus has been detected in the retinal periphery, it can be inspected by the fovea, where acuity is greatest. Figure 3.51 shows that there are far fewer cones away from the fovea so the ability to distinguish colour is substantially reduced away from the fovea.

Question 3.15

When the pupil is 2 mm in diameter, the diffraction pattern produced on the retina (see calculation in Section 3.7.1) is only a few receptors wide. Diffraction is not therefore a significant limitation on acuity. The advantages of a fairly narrow pupil such as this are that the depth of field is good and the effect of aberrations is minimized.

(a) If the pupil is smaller than this then spreading due to diffraction at the pupil will become a significant factor.

(b) If the pupil is larger than this then diffraction will be of no concern but other factors such as aberrations and depth of field will lead to loss of acuity.

Question 3.16

The shape of the CSF tells us a great deal about visual performance. For example, it tells us that we are most sensitive to a medium level of spatial detail (4–8 cycles degree^{-1}), that we are extremely sensitive to appropriate stimuli (a fraction of 1% contrast), that we are relatively insensitive to uniform luminance (zero spatial frequency), and that we have good acuity (50–60 cycles degree^{-1}). The shape of the CSF can also be related to the receptive fields of retinal ganglion cells. For example, the above properties can be related to receptive fields that pool responses over a small retinal area (giving good sensitivity and acuity) and that have discrete excitatory and inhibitory sub-regions (giving low sensitivity to uniform illumination). Moreover, changes in the shape of the CSF under different viewing conditions are also informative. For example, the CSF shifts to the left under dim illumination or peripheral viewing. This suggests that receptive fields are larger, and thus more sensitive but less able to resolve fine detail (high spatial frequencies), under these conditions.

Question 4.1

An individual simple cell would respond to a line or an edge of a particular size and orientation in one small region of the image. On the face of it, then, a simple cell could signal the presence of a particular visual feature, such as a vertical edge in the centre of the image. However, as is pointed out in Chapter 14 of the Reader, individual simple cells are not really selective enough to play this role unaided – each responds to a range of orientations and stimulus scales. Therefore an individual simple cell plays a role as part of the overall *pattern of response across many cells* that signals the spatial structure of each small region of the image.

Question 4.2

Brightness is the perceptual correlate of physical illumination; lightness is the perceptual correlate of surface reflectance. When the sun disappears behind a cloud, the illumination of the scene will decrease, but surface reflectance is unaffected. Therefore brightness will decrease, but lightness will remain unchanged (because of lightness constancy).

You should note that in the vision literature and on the World Wide Web these two terms are often used interchangeably and sometimes confused with each other. In particular, some people use brightness specifically for illumination, some for luminance, others use it for both indiscriminately.

Strictly speaking, lightness is the perceptual correlate of surface reflectance and is therefore the proper term to be used when describing how we see objects. Brightness is the perceptual correlate of light intensity and is therefore the proper term to be used when describing how we see light. These are the senses in which the two terms are used in this course. The distinction is very important when considering the problems that the visual system faces when trying to interpret the pattern of light in a retinal image. However, in many situations, the distinction between surfaces and light is not important – one simply wants to describe the light entering the eye or in the image, without considering how it got there or how it should be interpreted. In these situations, brightness is the default term.

Question 4.3

In the traditional UK system the red and green wires had roughly equal overall reflectance, so they would appear the same shade of grey in a black and white photograph. Also, these wires were easily confused by colour deficient people, with potentially fatal consequences. This problem was actually serious because red–green colour deficiencies are relatively common, especially among males. In the new system (brown = live, blue = neutral, green and yellow = earth), the wires are more easily distinguished in black and white photographs and also by colour deficient people. However, many people believe that the new system is not ideal either, because the particular choice of colours and patterns is not cognitively intuitive and therefore difficult to remember; most people would more naturally associate brown with earth, for example. This is a nice example of a situation where correction of a sensory problem has created a cognitive problem. You may well be able to think of a solution that avoids both kinds of problem.

Question 4.4

Both cues rely upon the same principle; the comparison of the same scene viewed from more than one position. The main difference is that motion parallax collects these views together over time (as the observer changes position), whereas binocular stereopsis collects two views simultaneously (one from each eye). Neither is really better in general, though each has advantages in particular contexts. One cost of stereopsis, for example is that it requires both eyes to face roughly forwards and this tends to restrict the field of view out to the side. Motion parallax, on the other hand, requires movement of the observer. This may be one reason why predatory animals tend to favour stereopsis (since their own movements might give them away to their prey), whereas prey animals tend to use motion parallax (since it is more important for them to spot potential predators over a wider visual range).

Question 4.5

There are many situations where 'common fate' does not seem to capture what is needed. For example, when you move towards a target its image expands, so that features of the image move away from, rather than towards, a common point. If you move your eyes, the whole image moves in the opposite direction, but not everything in the image should be grouped together. More generally consider the movement of a walking person, where different points on the body may be moving in opposite directions, one arm swinging forward while the other leg swings backwards for example. Although common fate may be useful in general, grouping by image movement also needs more subtle rules.

Question 4.6

The most obvious scheme would be to suggest that chairs can be recognized by a common physical structure, so that bottom-up processes would need to provide the sort of information that corresponds to important features like chair legs. The visual cortex seems well adapted to this, since its processes can pick up oriented lines and edges in images. However, there is a huge range of physical structures that people will accept as chairs, as can be verified by a visit to any modern furniture shop. It may well be more useful to think of chairs (and other objects) as being defined by their functional potential, rather than just their physical structure (i.e. a chair is anything you can sit on in a context where it might be appropriate to sit down). If this is the case, bottom-up processes need to provide rather different information (e.g. about flatness, rigidity, and so forth). The cortex may be well adapted to providing this type of information too, but in order to find out, we need to ask rather different questions about its processes.

Question 5.1

Although we have evolved to be particularly good at face recognition, the finer details of the process have to be learnt, as when growing children acquire a more holistic approach. The learning takes place with the facial features that surround us, so unfamiliar facial types present us with stimuli for which we have not practised discrimination.

Question 5.2

Skilled readers appear to do more 'whole shape' processing, in the same way as adults process faces. It is known that inverting a face changes the analysis to a feature-by-feature approach, so by analogy we might expect inverted words to be processed letter-by-letter. This would probably lead to phonological processing.

Question 5.3

Without carrying out an experiment it is not possible to state which are the quickest words to read. However, the following issues are relevant:

- Word frequency. Less common words will take longer.

- Whether the visual or phonological route is used. The word *sight* is a homophone (site) so might take longer, if the wrong word is located first, by sound. However, *sight* is not a very regularly spelled word, so its distinctive shape is probably recognized visually. *Eye* is another homophone (I), and for many people *poor* is pronounced like pour, paw and pore, so the word might take a long time to process by sound.

- Context effects. By the time the reader reaches the last word of the phrase, the preceding three will have given a degree of predictability, which will speed recognition.

Acknowledgements

Grateful acknowledgement is made to the following sources for permission to reproduce material in this book:

Cover

Copyright © Mehau Kulyk Science Photo Library.

Figures

Figure 1.4: Copyright © The National Portrait Gallery; *Figure 3.19*: Copyright © Science Photo Library; *Figure 3.2*: Yarbus, A. L. (1967) *Eye Movements and Vision*, Plenum Press. Copyright © 1967 Plenum Press; *Figure 3.33*: Reprinted from Lowenstein, O. and Loewenfeld, I. E. (1959) 'Influence of retinal adaptation upon the pupillary reflex to light in normal man', *American Journal of Ophthalmology*, **48**, p. 545. Copyright © 1959 with permission from Elsevier Science; *Figure 3.39*: Courtesy of Dr Brian Burton, Boston University; *Figures 3.46 and 3.47*: Barlow, H. B. and Mollon, J. D. (1982) *The Senses*, Cambridge University Press. Copyright © Cambridge University Press 1982; *Figure 3.50*: Van Nes, F. L. and Bouman, M. A. (1967) 'Spatial modulation transfer in the human eye', *Journal of the Optical Society of America*, **57**, p. 402, 1967, Optical Society of America; *Figure 5.8*: Copyright © J. C. Revy/Science Photo Library.

Every effort has been made to trace all copyright owners, but if any has been inadvertently overlooked, the publishers will be pleased to make the necessary arrangements at the first opportunity.

Glossary for Block 4

11-cis-retinal   The substance that, when attached to one of the opsin proteins, forms one of the four types of visual pigment molecules in rods and cones. It is formed by a series of transformations from vitamin A.

aberrations   Distortions in optical images produced by the optical systems that form them. Aberrations arise from a number of well known causes and are classified accordingly. *See, for example*, chromatic aberration and spherical aberration.

accommodation   The process whereby the eye adjusts its focal length so as to bring objects at different distances into sharp focus on the retina.

achromatic colour   An achromatic colour is one that is without hue. Black, white and all the grey colours are examples of achromatic colours.

adapting grating   A grating used to fatigue the cells that it activates by the process of adaptation and employed in psychophysical studies of spatial vision.

additive colour mixing   Additive colour mixing describes the process by which light from different primaries is superimposed to generate a particular colour. Additive colour mixing is used to produce images in television and on computer screens.

additive primary colours   The three colours of light (red, green and blue) that are used in image-reproduction systems (such as television) based upon additive colour mixing. This choice of primaries allows the largest gamut of colours to be reproduced.

Airy disc   The diffraction pattern produced by a circular aperture. It has a bright central spot and alternating dark/light rings.

akinetopsia   The inability to detect or interpret visual motion.

aliases   Spurious low spatial frequency components that arise from spatial frequencies in a stimulus that are too high for the sampling rate used.

all-trans-retinal   The substance formed during the visual process by the action of light on 11-*cis*-retinal.

amacrine cell   A type of nerve cell without axons whose cell bodies lie in the inner nuclear layer of the retina. They connect laterally in the inner plexiform layer with bipolar and ganglion cells.

anomalous trichromats   Individuals with three types of cone pigment but with one of the pigments having a slightly different spectral sensitivity to that found in a normal human observer. Anomalous trichromats have good colour vision but are unable to discriminate all the colours that a normal observer can.

area 17   The primary visual cortex, also known as the striate cortex or V1. The number is that given by Brodmann in his numerical cytoarchitectonic classification of the cerebral cortex.

articulatory suppression   The use of continuous speech to make it impossible to 'sound out' words while reading. Also called concurrent vocalization.

assimilation   The reduction of colour (or brightness) differences, or enhancement of colour similarity, between adjacent surfaces, caused when the surfaces are of particular sizes and arranged in a particular spatial configuration.

asymmetric matching experiment   A technique used in psychophysical experiments in which the observer must compare and judge the perceptual equality of two stimuli that are presented against different backgrounds or in different contexts.

binocular   Used to describe either a visual structure that processes information from both eyes or a part of the visual field that can be seen through both eyes.

binocular disparity   The difference in the relative position of the image of the same object in the two eyes. Provides the basis for binocular stereopsis, a depth cue.

binocular stereopsis   A depth cue based on binocular disparity.

bipolar cell   A nerve cell with two axons extending from opposite sides of the cell body. In the retina these interneurons connect photoreceptors with ganglion cells.

black body radiation   Electromagnetic radiation that is in thermal equilibrium with matter at a fixed temperature. Its name derives from the fact that its spectrum is identical to that which would be emitted by a completely black 'body' (object) at an appropriate temperature.

blind spot   *See* optic disc.

blobs   Areas of striate cortex with high levels of the enzyme cytochrome oxidase, which appear dark when stained for microscopic examination with a substance specific for the enzyme.

blue–yellow ganglion cell   A retinal ganglion cell with a receptive field that responds as if it had received signals of one sign for blue light and the opposite sign for yellow light.

bottom-up   A form of information processing in which the results are determined solely by the nature of the incoming stimulus. Synonymous with data driven, this is the converse of top-down or concept driven.

brightness   The perceptual correlate of illumination.

Brodmann's areas   The 47 separate zones of the cerebral cortex differentiated by the German neuroscientist Korbinian Brodmann on the basis of their visually distinct appearance.

calmodulin   A small regulatory calcium-binding protein found in almost all tissues.

carotenes   A group of pigments that are yellow, orange or red in colour, found in many plants, such as carrots, yellow maize and green vegetables. The best known example is *beta*-carotene, a precursor to vitamin A.

carotenoids   A family of yellow, orange or red pigments manufactured by bacteria, fungi and plants. It contains two sub-groups: carotenes and xanthophylls.

choroid   The middle layer in the eye, rich in blood vessels, which lies between the retina (the inner layer) and the sclera (the outer layer).

chroma   The degree to which a colour contains hue; equivalent terms are saturation and colourfulness. Vividness and purity are also sometimes used to describe the degree to which a colour differs from being achromatic. It is one of the three attributes used to specify a colour in the Munsell system, the others being hue and value.

chromatic aberration   The blurring of the image produced by a lens caused by the variation of refractive index with frequency, so e.g. blue light is focused more strongly than is red light. Consequently, images of objects illuminated with white light exhibit a rainbow-like blurring.

chromatic adaptation   The adjustment of visual sensitivity caused by selective exposure to particular wavelengths or bands of wavelengths of light.

chromatic colour   All colours are either chromatic or achromatic. Chromatic colours are those that possess hue.

chromaticity diagram   *See* CIE chromaticity diagram.

chromophore   A group of atoms in a molecule that gives colour to that substance. In the rod and cone visual pigments, 11-*cis*-retinal, which is bound to the protein opsin, is the chromophore and gives the pigment its colour.

CIE chromaticity diagram   The CIE chromaticity diagram is a two-dimensional map of colour space where colours of equal luminance are plotted. The chromaticity coordinates are derived from normalized tristimulus values.

CIE system   The CIE system refers to a method of colour specification that was introduced in 1931 by the Commission Internationale de l'Eclairage. It includes the colour-matching functions of the standard observer and definitions of standard illuminants and viewing geometries.

ciliary muscle   The muscle, contained in the ciliary body, which changes the shape of the lens when focusing. *See also* accommodation.

cilium   A cell organelle consisting of a characteristic organization of internal fibres or tubules. In rods and cones the outer and inner segments are connected by a ciliary bridge in which the characteristic organization has lost the central two tubules.

cognitive psychology   The branch of psychology concerned with recognition, understanding, memory and allied processes.

colour constancy   The phenomenon whereby the apparent colour of an object does not change even though the spectral distribution of the illuminating light, and hence of the reflected light, has changed.

colour contrast   The enhancement of colour differences between adjacent surfaces; typically, the effect caused when one surface induces its opponent colour in an adjacent surface, e.g. when a green background induces a reddish tint in a figure that it encloses.

colour space   A colour space is a two- or three-dimensional map used to plot colour stimuli. The CIE system is a very well known colour space that has many applications in applied colour science.

colourfulness   *See* chroma.

colour-matching functions   The amounts of each of three primaries that in an additive mixture produce a match to a monochromatic stimulus, determined experimentally over the range of the visible spectrum. The XYZ colour-matching functions of the CIE system form the basis of modern colorimetry.

colour opponency   *See* opponent processing.

common fate   In object recognition, the idea that individual image elements moving in the same direction should be grouped together.

complementary hue   Each hue has a complementary hue, that when the two are mixed additively, results in white or grey. Examples of complementary pairs are blue/yellow and red/cyan.

complex (cortical) cells   A type of cell in the primary visual cortex that responds to an appropriately oriented stimulus anywhere within its receptive field.

concept driven   *See* top-down.

concurrent vocalization   *See* articulatory suppression.

cone pedicle   Specialized synaptic terminal of cones consisting of many 'triads'. Each triad is composed of an invagination or infolding of the membrane that is occupied by two horizontal cell processes associated with a synaptic ribbon. Bipolar cells contact the pedicle through invaginating processes below the synaptic ribbon and also at the base of the pedicle.

cones   Visual photoreceptors in the retina that respond to normal light levels. In humans there are three spectral classes, with peak sensitivities to short-, medium- and long-wavelength light conferring trichromatic colour vision.

conjugate (version) movements   Synchronous identical movements of the two eyes.

connectionism   A school of thought which proposes that mental processes are implemented in large networks of interconnected neurons (neural nets), where the strengths of the interconnections (synapses) determine the behaviour of the network.

consensual light reflex   An alteration in the size of the pupil of one eye in response to a change in the light intensity on the other eye. This is an involuntary reflex.

constructive interference   Occurs when two waves superpose such that their combined displacements add together in phase. This results in a wave with a larger amplitude than either of the contributing waves.

contrast   The relative luminance of different points in the image.

contrast sensitivity   The reciprocal of the contrast threshold.

contrast sensitivity function (CSF)   The contrast sensitivity of the eye expressed as a function of spatial frequency.

contrast threshold   The minimum contrast at which an observer can reliably detect the presence of a grating. Contrast (modulation) is defined as $M = (I_{max} - I_{min})/(I_{max} + I_{min})$ where I_{max} represents the maximum intensity and I_{min} the minimum intensity.

convergence   The degree to which the eye must be rotated to bring the image of an object onto the fovea.

converging system   A system that increases the convergence (or reduces the divergence) of an incident wavefront. By convention, a converging lens always has a positive focal length.

cornea   A transparent layer of tissue that forms the front part of the eye, over the iris and lens. The cornea carries out most of the refraction of the light entering the eye, enabling it to be focused on the retina.

cyclic guanosine monophosphate (cGMP)   A cyclic nucleotide that in rods and cones controls the opening of cation channels in the outer segment. A drop in concentration leads to the closure of the channels.

cytoarchitectonics   The microscopic study of the distribution of neural cell types and organization within the brain.

dark adaptation   The changes that need to take place in the eye to enable objects to be seen clearly when moving from a brightly lit environment to a relatively dark one.

dark current   The flow of sodium ions into the outer segment, which occurs in photoreceptors in the dark. The sodium movement is balanced by an active sodium pump on the inner segment. Light stimulation causes a reduction in the dark current.

data driven   *See* bottom-up.

delayed inhibition   The inhibition of a retinal cell by a receptor cell only after a delay following the initial response to a stimulus. The phenomenon of delayed inhibition is the key to the function of a cell acting as a motion detector.

deoxygenated haemoglobin   Haemoglobin is the oxygen-carrying molecule within red blood cells. It is able to transport oxygen around the body because it can exist in two forms, oxygenated and deoxygenated. It is in the latter form when it has 'delivered' its oxygen load to, for example, a region of the brain. Of the two, only deoxygenated haemoglobin has suitable magnetic properties to give a magnetic resonance image.

depth of field   The range of distances within the object field over which an acceptably in-focus image is produced at a fixed image plane.

destructive interference   Occurs when two waves superpose such that their combined displacements add together out of phase. This results in a wave with a smaller amplitude than either of the contributing waves; when they are exactly out of phase, the resultant displacement is zero.

deuteranopes   Individuals who suffer from a form of dichromatic vision where the normal M-cone pigment is absent.

deuteranopia   Deuteranopia refers to a form of dichromatic vision where the normal M-cone pigment is absent.

dichromatic vision   The colour vision that is experienced by humans with only two classes of cones rather than the usual three. Dichromats are therefore colour-defective observers.

dichromats   Individuals with only two classes of cones rather than the normal three (usually lacking either L cones or M cones).

diffraction   The spreading or bending of waves as they pass through an aperture or round the edge of an obstacle.

diffuse bipolar cell   A bipolar cell located away from the fovea whose broad (diffuse) dendritic tree makes contact with from 5 to 14 cones.

dioptre (D)   A unit used to express the power of a lens and defined as $1\,D = 1\,m^{-1}$.

dioptric apparatus   The parts of the eye that contribute to the refraction of light to form an image on the retina.

diplopia   Double vision, caused by lack of coordination between the muscles that move the eyes to bring the image into correct alignment on both retinae.

direct light reflex   The reflex that causes the pupil of the eye to contract when light is shone on the eye. *See also* consensual light reflex.

direction selectivity   The extent to which cortical cells respond more to one direction of retinal motion than to others.

disjunctive (vergence) movements   Synchronous movements of the two eyes that are equal and opposite.

diverging system   A system that increases the divergence (or reduces the convergence) of an incident wavefront. By convention, a diverging lens always has a negative focal length.

dual route theory of reading   A theory about the process of reading which claims that printed word re analysed by two routes: visual and phonological.

dyslexia   A dysfunction in the ability to read, colloquially known as 'word blindness'.

electromagnetic wave   A wave that consists n oscillating electric field and an associated magnetic field oscillating at the same frequency. The two fields are mutually perpendicular and perpendicular to the direction of propagation of the wave. Electromagnetic waves travel through a vacuum at the speed of light.

emotional Stroop test   A form of the Stroop effect, in which slow responses result from the use of words that have emotional significance to the reader.

end-stopped cells   An alternative (more modern) name for a hypercomplex cell.

far point   The most distant point on which the eye can clearly focus. For a 'normal' eye, it is at infinity (i.e. rays from this point are parallel as they enter the eye).

feature detectors   The notion that an individual cell might signal the presence of a particular feature, such as an oriented bar or edge, in an image.

fixation axis   The line between the centre of rotation of the eye and the point of fixation.

focal length   The distance between the optical centre of a lens and its focal point. For a converging lens, it is the distance over which parallel rays are brought to a focus. By convention, the focal length of a converging lens is positive and that of a diverging lens is negative.

focal plane   The plane that passes through a lens's focal point and is perpendicular to its optical axis.

focal point   The point to which parallel rays striking a converging lens converge, or from which parallel rays striking a diverging lens appear to diverge.

fovea   A depression or pit in the centre of the retina containing densely packed cones and no overlying interneurons that is specialized for high visual resolution.

frontal lobe   One of the four lobes of the cerebral cortex situated at the front of the brain. It is associated with motor control and higher mental processes.

functional imaging techniques   The use of imaging techniques such as MEG, EEG, PET and fMRI to locate distinct anatomical regions of the brain that perform particular sensory, motor or cognitive functions in response to a carefully controlled stimulus.

fusiform gyrus   An area of cortex involved in face recognition.

gamut   The gamut of a colour-reproduction device is the range of colours that can be physically reproduced by that device. Most colour-reproduction devices have a gamut that is no more than 50 per cent of all the possible colours that could be reproduced.

ganglion cell layer   The layer in the retina where the retinal ganglion cells are located. It is the first layer that incident light meets and is next to the inner plexiform layer.

ganglion cell   Nerve cells located in the retina whose axons form the optic nerve, which connects to the lateral geniculate nucleus in the middle of the brain. They receive their inputs from the photoreceptors (rods and cones) via intermediate horizontal, amacrine and bipolar cells.

geniculostriate pathway   The neural pathway from the retina, via the lateral geniculate nucleus, to the primary visual cortex.

Gestalt psychology   The holistic approach to perception developed by a group of psychologists who believed that 'the whole is greater than the parts' and that context is extremely important in understanding perception. Gestalt psychology began in Germany prior to World War I and became widespread there during the 1920s and 1930s.

Grassman's additivity law   Grassman's law states that stimuli of the same colour produce identical effects in mixtures regardless of their spectral composition.

grouping   Deciding which parts of an image belong together, when combining the information resulting from the separate sub-modalities during the later stages of visual processing.

guanosine diphosphate (GDP)   A nucleotide that when further phosphorylated becomes a high energy molecule, guanosine triphosphate (similar to ATP). It is involved in the energy-requiring activation of transducin.

guanosine triphosphate (GTP)   A high energy molecule (similar to ATP) that is involved in energy-requiring processes such as the activation of transducin in the visual system.

haploscopically   Term used to describe the presentation of two images side-by-side to an observer so that each eye views just one of the images, a technique used in colour matching experiments.

higher-level cognitive functions   The more complex 'thinking' processes of the brain.

homophone   A word with the same pronunciation as another, but with different spelling.

horizontal cell   A nerve cell whose fibres extend laterally (horizontally) located in the outer plexiform layer of the retina. They are involved in lateral inhibition.

hue   The quality of a colour that determines whether it is red, green, blue, etc. It is one of the three attributes used to specify a colour in the Munsell system, the others being chroma and value.

human homologue   When a region of the brain is first identified in an animal (such as the Macaque monkey), and then a region with similar properties is discovered in humans, this area is commonly described as the human homologue of the region found in the animal.

Huygens' principle   A principle asserting that each point on a wavefront can be regarded as a source of secondary waves. The secondary waves from each point spread out equally in all directions. The position of the new wavefront after a brief time interval is the envelope of these individual waves (a line, curve or surface drawn tangentially through the individual secondary wave-fronts). This in turn acts as a source of secondary waves, thus accounting for the continued propagation of the wave.

hypercolumn   The set of cells in the primary visual cortex that processes one small region of the retinal image.

hypercomplex (cortical) cell   A type of cell in the primary visual cortex with a receptive field similar to a complex cell, but that responds only if an appropriately oriented stimulus ends somewhere within the receptive field.

hyperopia A defect of vision commonly known as long-sightedness in which the near point of the eye is considerably further away than normal. The eyeball is too short, or the refracting power of the eye is too weak to focus on near objects, and light from an object located at the normal near point would be focused behind the retina. Long-sightedness is also known as hypermetropia.

hyperpolarization An increase in the potential difference (voltage) across the cell membrane making the interior of the cell more negative than at rest. Rods and cones hyperpolarize on activation.

ideogram/ideograph A symbol (as opposed to a printed word) that represents an object or concept.

illuminance The amount of light falling upon a surface.

inner nuclear layer A layer in the retina consisting of the cell bodies of bipolar, horizontal and amacrine cells.

inner plexiform layer A layer in the retina composed of the synaptic connections of bipolar, ganglion and amacrine cells.

intercalated (I) layer One of the set of layers lying between the P and M layers of the LGN, also called koniocellular (K) layers. They contain a morphologically and physiologically distinct class of neurons.

interneuron Nerve cell connecting two or more other neurons. In the retina these link photoreceptors with ganglion cells.

invagination The pocket or pockets at the base of rods and cones (synaptic region) that partially enclose the processes of horizontal and bipolar cells.

iodopsin The visual pigment of the three types of cone. They each consist of a protein, opsin, to which is attached a molecule of 11-*cis*-retinal. Opsin is a member of the 7- transmembrane family, so called because the protein chain traverses the lipid bilayer of the membrane seven times. The 11-*cis*-retinal portion is the chromophore in all three types of iodopsin; the different absorption characteristics arise because they each contain a slightly different opsin.

iris The adjustable diaphragm in the human eye, located just in front of the crystalline lens, which gives the eye its characteristic colour. The hole at the centre of the iris, through which light enters the eye, is the pupil.

isomers Compounds with the same molecular formulae (i.e. that contain the same number and type of atoms) but which have different shape molecules that do not easily interconvert.

koniocellular (K) layer *See* intercalated (I) layer.

laser in-situ keratomileusis (LASIK) A surgical technique used to effect a refractive correction to the eye. A thin flap of cornea is lifted and some of the central stroma is removed with a laser such that when the flap is replaced, the corneal shape is changed.

lateral geniculate nucleus (LGN) One of a pair of nuclei in the thalamus to which retinal ganglion cells project and which relays visual information to the primary visual cortex.

lateral inhibition The output neurons of the retina (ganglion cells) are connected to the photoreceptors (rods and cones) by interneurons (horizontal, bipolar and amacrine cells). The connections are such that each ganglion cell receives two types of information: excitatory from a particular photoreceptor and inhibitory from adjacent receptors. This latter effect is lateral inhibition.

laws of Prägnanz Basic rules describing visual grouping processes compiled originally by the Gestalt psychologists during the 1920s and 1930s.

least distance of distinct vision *See* near point.

lens A device (usually a specially shaped piece of glass, or other transparent material with a refractive index different from that of its surroundings) that is able to make incident parallel rays converge to a point or appear to have diverged from a point.

lens equation The equation

$$\frac{1}{u} + \frac{1}{v} = \frac{1}{f}$$

that relates the object distance u, the image distance v and the focal length f of a (thin) lens.

lexical decision task The task of deciding whether a string of letters is a correctly spelled word or not.

lexical entry An entry (i.e. a word) in the lexicon.

lexicon The mental store of words known to the user, presumed to contain meanings, pronunciations and spellings.

LGN *See* lateral geniculate nucleus.

lightness The perceptual correlate of surface reflectance.

lightness constancy The perceptual phenomenon by which the perceived reflectance (i.e. lightness) of surfaces remains constant despite variations in illumination.

lightness contrast   The phenomenon by which perceived surface reflectance (i.e. lightness) depends upon the background lightness (e.g. things appear lighter on a dark background).

lipid bilayer   Lipids (e.g. phosphoglycerides) have a polar head and a hydrophobic tail. In an aqueous medium they form a bilayer with their hydrophobic tails inside and their hydrophilic polar heads pointing outwards. A lipid bilayer constitutes the basic structure of biological cell membranes.

low-level visual attribute   Simple visual features that can be processed in early visual areas. One example is the detection of edges in a visual scene.

luminance   The amount of light reflected from a surface.

lysine   A naturally occurring amino acid with a side chain containing an amino group ($-NH_2$). 11-*cis*-retinal attaches to opsin in rhodopsin and iodopsin by linking with the side chain amino group of a lysine residue.

macula lutea   A 3–4 mm diameter central region of the retina, including the fovea and often referred to as the 'yellow spot', which contains a yellow screening pigment consisting of a mixture of carotenoids.

magnocellular (M) layer   The two lower or ventral layers of the six main layers of the LGN, composed of neurons with larger cell bodies than those in the four dorsal layers. This region receives input from the retina primarily from parasol ganglion cells (M cells). The term is also used to describe the retinal ganglion cells that project to the magnocellular layers of the LGN.

M cell   *See* parasol ganglion cell.

mental rotation   The process which enables objects presented visually in one orientation to be visualized in another orientation.

midget bipolar cell   A bipolar cell in the central retina (fovea) that receives input from one (or occasionally two) cones and connects with a single midget ganglion cell.

midget ganglion cell   Ganglion cell (P cell) identified in the foveal part of the retina. It receives input from one (or occasionally two) cone photoreceptors through a single midget bipolar cell. Their axons connect to the parvocellular layers of the LGN.

minutes of arc   An angular measure equal to 1/60th of a degree.

modulation   The variation between the darkest and lightest regions of a pattern. It is defined as $M = (I_{max} - I_{min})/(I_{max} + I_{min})$.

modulation transfer function (mtf)   The modulation transfer function describes how well an imaging system transfers modulation from object to image. At each spatial frequency the mtf is defined by:

$$\text{mtf} = \frac{\text{modulation in image at a particular spatial frequency}}{\text{modulation in object at the same spatial frequency}}$$

monochromatic   A term used to describe electromagnetic radiation of a single frequency.

monochromats   Individuals who possess only one class of cone photoreceptor and cannot perform any colour discrimination. They perceive the world only in shades of grey. Monochromats are extremely rare in the human population.

motion parallax   The difference in the motions of points at different positions in an image when the observer changes position. Since the speed of retinal motion depends on the distance of the object, this is a depth cue.

Munsell colour-order system   The Munsell colour-order system describes a collection of coloured chips or patches that are arranged in a three-dimensional manner and used for colour communication. This system was designed by the artist Albert Munsell.

myopia   A defect of vision commonly known as short-sightedness in which the far point of the eye is considerably closer than normal. The eyeball is too long, or the refracting power of the eye is too great to focus on distant objects, and light from such an object would be focused in front of the retina.

nasal   The half of the retina that is closer to the nose.

near point   The closest point on which the eye can focus. It is conventional to assume a standard near-point distance of 250 mm for a 'normal' eye.

near reflex   The decrease in diameter of the pupil when the eyes converge to look at a close object. This is a reflex, i.e. an involuntary change.

Necker cube   An 'open' cube, in which all edges are visible and are represented by lines. The layout is such that the orientation of the cube remains ambiguous and appears to alternate between possibilities.

neural adaptation   The increase in sensitivity of the retina by the dynamic reorganization of ganglion cell receptive fields so that the inhibitory sub-regions are less powerful. This has the effect of increasing sensitivity at the expense of spatial resolution.

neural image   The retinotopic pattern of neural activity produced by the stimulated photoreceptors, which directly mirrors the pattern of light in an image on the retina.

neural nets   *See* neural network.

neural network   A network of neurons, or more often a computer representation of such a network, implemented to test connectionist ideas. *See also* connectionism.

neuropsychology   The branch of psychology that seeks to deduce how the normal brain functions by observing the nature and degree of dysfunction in brain-damaged patients.

neurotransmitter   A chemical messenger released at a synapse by a presynaptic nerve terminal that interacts with receptor molecules on the postsynaptic membrane. In the case of rods and cones, for example, this is glutamate (an amino acid).

occipital lobe   One of the four lobes of the cerebral cortex situated at the rear of the brain. It is responsible for visual processing.

ocular dominance   The extent to which the input to an individual cell in the primary visual cortex from one eye is stronger than that from the other.

OFF-centre   Cells, such as bipolar and ganglion cells in the retina, whose receptive field centres are inhibited by stimulation.

ON-centre   Cells, such as bipolar and ganglion cells in the retina, whose receptive field centres are excited by stimulation.

opponent processing   Opponent processing refers to the fact that the visual system combines the responses of the L, M and S cones to generate three new signals that encode luminance, redness–greenness, and yellowness–blueness.

opsin   The protein part of a visual pigment molecule consisting of a chain of about 350 amino acid residues.

optic chiasm   The point in the visual pathway where the two optic nerves meet and half the fibres in each optic nerve cross over to the other side of the brain. This ensures that the images of the same object in the left and right eyes are processed in the same hemisphere.

optic disc   The area in each eye, commonly called the blind spot, where the axons of the retinal ganglion cells forming the optic nerve come together and leave the eye. There are no photoreceptors in this area.

optic nerve   The nerve that carries all the visual signals from the retina to the lateral geniculate nucleus in the brain. It consists of the axons of retinal ganglion cells, of which there are about 1 000 000.

optic radiation   Pathway carrying visual signals from the lateral geniculate nucleus to the primary visual cortex.

optic tract   Formed from half the fibres in each optic nerve, each optic tract carries all the visual information relating to the opposite (contralateral) side of the visual field. Nerve fibres from the nasal side of the retina of each eye cross over to join fibres from the temporal side of the retina of the opposite eye.

outer nuclear layer   A layer of the retina consisting of the cell bodies of rods and cones.

outer plexiform layer   A layer of the retina composed of the synaptic connections of rods and cones, bipolar and horizontal cells.

pandemonium model   A feature-based object recognition scheme involving a population of 'demons' each with a particular task. Feature demons signal the presence of features in an image; cognitive demons look for feature combinations; finally, a decision demon selects the most active cognitive demons.

parallel distributed processing (PDP)   Used in the context of connectionism, the term reflects the simultaneous activation that occurs in highly interconnected neural nets.

parallel processing   The simultaneous processing of information.

parasol ganglion cell   Ganglion cell (M cell) with a relatively large dendritic field connecting to a relatively large number of diffuse bipolar cells. Their axons connect to the magnetocellular layers of the LGN.

parvocellular (P) layer   The four upper or dorsal layers of the six main layers of the LGN, composed of neurons with relatively small cell bodies. This region receives input from the retina primarily from midget ganglion cells (P cells). The term is also used to describe the retinal ganglion cells that project to the parvocellular layers of the LGN.

P cell   *See* midget ganglion cell.

pentachromats   Pentachromats are creatures whose colour vision is based upon five basic types of receptor. Some pigeons and ducks are pentachromats.

phagocytosis   The uptake by a cell of particles or cells into cytoplasmic vacuoles.

phonological dyslexia   A form of dyslexia that renders a patient unable to derive the pronunciations of words (or nonwords) from their spelling, although words can potentially be recognized via the visual route.

phonological route   An information processing pathway that accesses a word's meaning by first deriving its pronunciation from the spelling.

phosphodiesterase (PDE)   A cytoplasmic enzyme that hydrolyses (degrades) a cyclic nucleotide. In visual transduction PDE hydrolyses cyclic guanosine monophosphate (cGMP) to guanosine monophosphate.

photons   A particle (or quantum) of electromagnetic radiation. A photon of frequency f carries an amount of energy given by $E = hf$ where h is Planck's constant.

photopic   The type of vision that occurs when the level of illumination is high, which is usually mediated by cones. In humans, photopic vision is trichromatic because of the three spectral classes of cone.

photopigment   The light-sensitive pigments found in the retina of humans and other animals. The human retina contains two main classes of photopigment: rhodopsin (in rods) and iodopsin (in cones).

photoreceptor   Sensory cells specialized for the detection of light. The principal photoreceptors of the retina are rods and cones.

photorefractive keratectomy (PRK)   A surgical technique whereby the surface of the corneal stroma is reshaped using a laser to effect a change in the refractive power of the cornea.

polyene chain   A carbon chain consisting of alternating single and double bonds.

polypeptide   A linked sequence of amino acids, the units that form the basic building blocks of all proteins.

Ponzo illusion   When two horizontal rods of equal length are drawn over an upside down V the upper rod appears longer than the lower rod.

population code   The way that certain visual stimulus attributes (e.g. orientation) are encoded not by individual cortical cells but by the pattern of activity across a whole ensemble or population.

power   In optics, the power of a lens is a measure of its ability to refract light. It is defined as the reciprocal of the focal length of the lens (i.e. $P = 1/f$). If f is expressed in metres, then P will be measured in dioptres. By convention, a converging lens has a positive power and a diverging lens has a negative power.

presbyopia   A defect of vision, common in older people, caused by the eye muscles weakening and the eye lens tissues stiffening. The most noticeable effect is a reduction in the range of accommodation, particularly the loss of close vision. *See also* myopia and hyperopia.

primaries   A primary is a component of an additive or subtractive colour mixing system. The commonly used additive primaries are red, green and blue (RGB), but for precise colour specification, the CIE XYZ colour-matching functions derived from the RGB primaries are used instead. The subtractive primaries are cyan, magenta and yellow.

primary visual cortex   The part of the cortex that receives the most direct projection of visual information from the eyes via the lateral geniculate nucleus. It is also called the striate cortex, area 17, and (in primates only) V1.

principle of superposition   The principle of superposition states that if two or more waves meet at a point in space, then at each instant of time the net disturbance at that point is given by the sum of the disturbances created by each of the waves individually.

principle of univariance   Absorption of a quantum of short wavelength (e.g. 400 nm) light by a photoreceptor causes the same qualitative response as the absorption of a quantum of long wavelength (e.g. 700 nm) light despite the fact that the former is of higher energy. Consequently, any single photopigment does not encode any information about the relative spectral composition of the light, only its rate of absorption.

prosopagnosia   A condition resulting from brain damage, in which the sufferer is unable to recognize familiar faces.

protanopes   Individuals who suffer from a form of dichromatic vision where the normal L-cone pigment is absent.

protanopia   A form of dichromatic vision where the normal L-cone pigment is absent.

pseudohomophone   A nonword, with pronunciation identical to that of a real word (e.g. brane).

psychometric function   A plot of response versus the intensity of a stimulus.

pupil   The circular opening in the centre of the iris, the size of which regulates the amount of light that reaches the retina. In a typical human eye, the pupil diameter can vary from about 2 mm (in bright light conditions) to about 8 mm.

pursuit movements   Movements that allow the eye to pursue moving objects and maintain the image of the object in the same position on the retina.

Rayleigh scattering   Scattering of light from particles very much smaller than its wavelength (e.g. molecules). The scattering intensity is proportional to the fourth power of the frequency, so blue light scatters about ten times as much as red light. This accounts for the blue colour of the sky and the red colour of sunsets.

receptive field   A fundamental property of sensory neurons; the region of sensory space which, when stimulated, evokes a response from an individual neuron. In the visual system, the area of visual space within which patterns of light will elicit a response.

recognition acuity   The ability to recognize an object (e.g. a letter).

refraction   The process whereby the direction of propagation of a wave passing from one medium to another is changed as a result of its change of speed. *See also* Snell's law of refraction.

refractive index   The refractive index of a medium is the ratio of the speed of light in a vacuum to the speed of light in the medium. Refractive index varies with wavelength and is always greater than one.

refractive power   The ability of a curved surface to refract light. It can be calculated using the equation $P = (n_2 - n_1)/r$ where P is the refractive power in dioptres, n_1 and n_2 are the refractive indices of the first and second media, and r is the radius of curvature of the surface (in metres).

relative reflectance   The proportion of the illumination falling on a surface that a surface reflects.

resolution acuity   The minimum separation required for a subject to resolve two adjacent discrete elements of a pattern.

reticular formation   Part of the midbrain containing interconnected networks of neurons associated with control of alertness and receiving inputs from several sensory systems.

retina   The neural layer lining the back of the eye, specialized for the detection and transduction of light. Neural information from the retina is transmitted to the brain via the optic nerve.

retinal   The aldehyde derived from vitamin A (retinol) which combines with opsin to form a visual pigment.

retinal ganglion cells   The 'output' stage of the retina and the first point at which action potentials are generated. The axons of the retinal ganglion cells are the individual fibres of the optic nerve.

retinal pigment epithelium (RPE)   Black pigmented layer lining the back of the eye behind the retina. The cells of the RPE interdigitate with the rods and cones and serve a number of functions including the absorption of stray light, the photoisomerization of retinal and the phagocytosis of the tips of the photoreceptor outer segments.

retinotopic flatmap   Retinotopic data obtained from imaging the visual cortex (using fMRI, for example) can be difficult to interpret. This difficulty arises because the grey matter of the cortex is a two-dimensional sheet which is folded in a highly convoluted way. Using software, this can be unfolded to form what is known as a 'flatmap', making it much easier to interpret retinotopic data.

retinotopic map   The organization of certain higher areas of the visual system (e.g. the LGN and V1) such that adjacent sets of neurons process the signals from stimuli that are adjacent to each other in the visual world. In this way the spatial pattern of the retinal image is preserved.

retinotopic mapping   Determining the boundaries of the visual areas by systematically and sequentially stimulating the different parts of the visual field while imaging the brain using fMRI.

retinotopic organization   *See* visuotopic representation.

retinotopic representation   *See* visuotopic representation.

rhodopsin   The visual pigment of rods consisting of a protein, opsin, to which is attached a molecule of 11-*cis*-retinal. Opsin is a member of the 7- transmembrane family, so called because the protein chain traverses the lipid bilayer of the membrane seven times. The 11-*cis*-retinal portion is the chromophore of rhodopsin.

rods   Highly sensitive visual photoreceptors specialized for the type of vision that occurs in dim light (scotopic or 'night vision'). In humans there is one spectral class so that rod vision is monochromatic.

rod spherule   Specialized synaptic terminal of rods composed of an invagination or infolding of the membrane that is occupied by the processes of two horizontal cells and two or more rod bipolar cells associated with a synaptic ribbon.

saccades   Brief rapid movements of the eye between fixation points.

saccadic omission   The lack of awareness of any blur during saccades. This is thought to be due to masking by the stimuli perceived just before and just after the eye movement.

saturation   *See* chroma.

sclera   Colloquially called the 'white' of the eye, the sclera forms most of the outer coat of the eyeball. Its function is to protect the internal contents and maintain the shape of the eyeball.

scotopic   The type of vision that occurs in dim light ('night vision'), which is mediated by rods. Since there is only one spectral class of rod photoreceptors, scotopic vision is monochromatic.

segmentation   Deciding which parts of an image belong to separate objects, when combining the information resulting from the separate sub-modalities during the later stages of visual processing.

simple (cortical) cells   A type of cell in the primary visual cortex with a receptive field that responds to a stimulus at a specific position with an appropriate orientation.

sine-wave grating   A grating whose intensity varies from white to black in the form of a sine wave. The rate of change is determined by the spatial frequency.

size constancy   The perception of an object to be a particular size irrespective of the absolute size of its image on the retina.

Snell's law of refraction   A law stating that when a wave travels from one medium in which its speed is v_1 to another in which its speed is v_2, the angle of incidence i and the angle of refraction r are related by:

$$\frac{\sin i}{\sin r} = \frac{v_1}{v_2} = \frac{n_2}{n_1}$$

where n_1 and n_2 are the refractive indices of the two media.

spatial antagonism   The phenomenon that the firing of one cell in the human visual system can be inhibited by the firing of a neighbouring cell.

spatial frequency   The frequency of change across an image or a grating. High spatial frequencies include those changes that occur in very close proximity, such as fine lines, low spatial frequencies include those changes which occur over greater distances, such as broad bands. The units for spatial frequency are length^{-1} or, in the case of the eye, degree^{-1}.

spectral antagonism   The phenomenon that the responses of cells responding primarily to one part of the visible spectrum can be combined with opposite sign with the responses of cells responding to another part of the visible spectrum. This phenomenon results in opponent processing.

spectral distribution   The amount of light emitted by a source expressed as a function of wavelength.

spectral power distribution   *See* spectral distribution.

spectral reflectance   The property of a surface that gives rise to its colour, this is the fraction of light reflected as a function of wavelength in the visible spectrum.

specular reflection   Mirror-like reflection from shiny surfaces. The opposite of diffuse reflection (from matt surfaces), where the light is reflected in all directions evenly.

spelling-to-sound correspondence rules   The phonetic rules that govern the relationship between individual letters or small groups of letters and their sound when pronounced. Words that conform to these rules are called regular, and those that do not are termed irregular.

spherical aberration   The blurring of the image produced by a lens with spherical surfaces, caused by the fact that rays further away from the optical axis are focused more strongly than rays close to the axis.

striate cortex   An alternative name for the primary visual cortex, derived from the characteristically striped appearance when examined microscopically.

stroma   The corneal stroma is the central layer and the main framework of the cornea, which gives the tissue its strength and rigidity.

Stroop effect   The effect of confusion, and hence slowness, which occurs when a person attempts rapidly to name the ink colours used to print incompatible colour words (e.g. the word 'blue' printed with red ink).

structural imaging techniques   Techniques such as CT and MRI that provide high resolution images of the internal structure of the body, particularly the brain. These techniques have important clinical applications, for example in the detection of tumours and other abnormalities.

subtractive colour mixing   The process by which dyes and pigments are physically mixed together to generate a subtractive mixture colour. Examples of subtractive colour mixing can be seen in paint systems and most printers.

surface dyslexia   A form of dyslexia in which the sufferer is unable to recognize words via the visual route, resulting in an ability to read only regular words (and nonwords), but not irregular words, such as 'colonel'.

surface spectral reflectance function   The function describing the proportion of incident light at each wavelength that a particular surface reflects. This is a constant property of a surface, determined by the material from which it is made.

synaptic cleft   The space separating nerve cells at a synapse, across which neurotransmitter must diffuse. In rods and cones the synaptic clefts are modified into the invaginations and triads of rod spherules and cone pedicles.

tachistoscope   A device for presenting visual stimuli for brief, controlled lengths of time.

temporal   The half of the retina that is further from the nose.

temporal frequencies   The frequency of change per unit time. Temporal frequencies are expressed in units of Hz (cycles per second).

tetrachromats   Tetrachromats are creatures whose colour vision is based upon four types of cone photoreceptor. There is some evidence that a small proportion of the female human population is tetrachromatic.

Thatcher illusion   The illusion, when viewing an upside-down picture of a face with certain features (e.g. the eyes or mouth) inverted, that the face is upright. This is because, when examining the picture, the process of mental rotation is carried out feature by feature, as

required. Because rotation is not a conscious process, the viewer is unaware of the differences in processing. The distortion is only apparent when the whole picture is turned upside down.

tilt after-effect (TAE)   The perception that, following adaptation by a grating tilted slightly away from the vertical, a vertical grating appears tilted in the opposite direction. This is the result of fatiguing the population of cells coding for the original grating orientation.

time-to-contact (TTC)   The time needed for a moving entity (e.g. a car) to reach a particular goal.

top-down   A form of information processing in which the outcome is strongly influenced by previously held concepts or experience. Synonymous with concept driven, this is the converse of bottom-up or data driven.

transducin   A G-protein (GDP-binding protein) found in the outer segments of rods and cones and involved in transduction.

transduction   A general term for the conversion of one form of energy into another. In vision, the process by which light energy in the form of absorbed photons is converted into a neural signal.

trichromatic vision   The normal colour vision that is experienced by humans. The term is sometimes used more generally to refer to the vision experienced by any creature with three types of cone photoreceptors.

tristimulus value   The amount of the additive primaries required to match a particular stimulus. The primaries may be R, G and B or the CIE colour matching functions X, Y and Z.

two-interval forced choice technique (2IFC)   A psychophysical technique widely used in vision science. A single trial consists of two brief stimulus presentations separated in time, each signalled by an auditory beep. One of the intervals, chosen at random, contains the test stimulus and the other contains no stimulus. The observer has to decide which interval contains the test stimulus and indicate their response by pressing one of two buttons. By performing many trials it is possible to generate a psychometric function.

V1   An alternative name for the primary visual cortex.

value   One of the three variables used to specify a colour in the Munsell system (the others being hue and chroma), it denotes brightness or lightness. The scale of value ranges from 0 (black) to 10 (white).

vergence movements   *See* disjunctive movements.

vernier acuity   The ability to detect whether two lines, laid end to end, are continuous or offset.

version movements   *See* conjugate movements.

vestibulo-ocular movements   The movements of the eye that maintain the direction of fixation while the head is rotated.

visual acuity   The ability of the eye to distinguish between objects that lie close together, sometimes called sharpness of vision. Visual acuity depends on the ability of the eye to focus incoming light to form a sharp image on the retina, and on the size of receptive fields in the retina.

visual angle   The angle subtended by an object at the pupil of the eye.

visual psychophysics   The study and measurement of the performance of the human visual system. This is done by showing subjects visual stimuli (usually close to the limits of detectability) and measuring how well they can perceive those stimuli.

visual route   The information processing path used in reading, which accesses a word's meaning by recognizing the overall pattern of its printed form.

visual sub-modality   One particular aspect of the visual sensation, e.g. colour, spatial structure or motion.

visuotopic representation   Neurons within the visual cortex with a defined map of a visual attribute.

wavefront   A line (in two dimensions) or a surface (in three dimensions) that connects points in a wave that have the same phase.

white light   Electromagnetic radiation that contains all the wavelengths in the range 380–780 nm (the visible region) and results in the sensation of 'white'.

wide-field bipolar cell   A bipolar cell located towards the periphery of the retina with a wide dendritic tree that makes contact with a large number of cones (greater than 15–20).

word superiority effect   A phenomenon in which tachistoscopically-presented letters are more accurately reported when they form part of a word than when they are displayed individually.

X-ray diffraction pattern   The pattern produced on an X-ray sensitive detector when a beam of X-rays is scattered (diffracted) by regular atomic-scale structures within a substance. Constructive and destructive interference give rise to light regions (diffraction maxima) and dark regions (diffraction minima) respectively on the detector. The relative positions of the diffraction maxima allow the detailed molecular structure of the substance to be determined.

zonular fibres   The ligaments that connect the edge of the lens in the eye to the muscular ciliary body.

Index

Entries and page numbers in **bold type** refer to key words which are printed in **bold** in the text and which are defined in the Glossary.

A

aberrations (lenses) **22**, 53–4
accommodation 31
additive primary colours 13
Airy disc 55, *56*
amacrine cells *36*, *37*, 42
area 17 of cerebral cortex **73**

B

binocular disparity 85
binocular stereopsis 6, **84**, 85–6
bipolar cells *36*, *37*, 42
black body radiation 9
blind spot *66*, 72
blinking *see* eyeblinks
blobs 82
bottom-up processing **71**, 91, 99
bright conditions *see* photopic conditions
brightness 80

C

carotenes 40
choroid *21*, **52**
chroma 13
chromatic aberrations 54
chromophores 39
CIE chromaticity diagram 13
ciliary muscle 21, 30
cognitive psychology 100
colour 11
colour constancy 7, *8*, **83**
colour opponency 82
colour science 13–14
colour vision 7, 43, 44–5, 82–3
  deficiencies 45
  luminance threshold for 51
common fate 87
complex cells 78
concept driven 99
cones (photoreceptors) **36**, *37*, 44–5
  distribution in retina 62, 67
  photochemistry of vision in 38–40, 57

conjugate movements of eyes **46**
connectionist theories **100**–1
consensual light reflex 22
constructive interference 35
context effects 99
contrast 63
contrast sensitivity 64
contrast sensitivity function (**CSF**) **62**–5, *67*, 79
contrast threshold 64
convergence 22, **46**, 49, 83
converging system (lens) **26**, 27, *29*, 32
cornea 21, *28*, 29, 32, 53
CSF *see* contrast sensitivity function

D

dark adaptation 56
dark current 40
data driven 99
depolarization of action potential 37, 40
depth of field 22
  and pupil size 52–3
depth perception 6
  relative 83–6
destructive interference 35
diffraction 32, **34**–5, 54–5, *56*
dim conditions *see* scotopic conditions
dioptre (**D**) (unit) 23, **28**
dioptric apparatus (of eye) **21**, 23–36
  as camera 23
  focusing the lens 29–31
  focusing power 23–9, 30
  refractive defects 31–2
diplopia 49
direct light reflex 22, 51
directionally-selective cells 84
direction selectivity 84
disjunctive movements of eyes **46**, 49
diverging system (lens) **27**, 32
double vision *see* diplopia
dyslexia 101

E

electromagnetic waves 9–10
end-stopped cells 78
eyeblinks 49
eyes 5
  adaptation to luminance conditions 50–57
  modulation transfer function 63, *64*
  movements 45–9
  refractive defects 31–2
  structure 21
  see also dioptric apparatus

F

facial recognition 95–7, 102
far point 29
feature detectors 78
fixation axes (eyes) **46**
focal length of lens **27**–8, 30
focal plane 27
focal point 27, *29*
Fourier analysis 15, 16
fovea *21*, 43
  cortical representation 73
fusiform gyrus 96

G

geniculostriate pathway 72
Gestalt psychology 87
'grandmother cell' concept 94, 100
gratings 16, *17*
Gregory, Richard 99
grouping 86–8, 90

H

horizontal cells *36*, *37*
hue 13
Huygens' principle 24, 34
hypercolumns (in cortex) **74**, 79, 82, 87–8
hypercomplex cells 78
hyperopia 31, 32

hyperpolarization (action potential) 37, 40

I

illuminance 79–80, 89
illusions *see* visual illusions
'impossible figures' 99
interneurons (retinal) 44, 72, 76
iodopsin 38, 40
iris 21, 22
isomerism *cis/trans* 38–9, 40
isomers 38

L

lateral geniculate nucleus (**LGN**) **72**–3, 82
laws of Prägnanz 87
learning 100, 101–2
least distance of distinct vision 29
lens (eye) **21**, 23, 27, 28–9
  aberrations 22
  focusing 29–31
lens equation 30
lenses 23, 26–8
  aberrations 53–4
  corrective 32
lexical decision task 94–5
LGN *see* lateral geniculate nucleus
light 9–11
  reflected 11, 13
  spatial variation of intensity 15–18
  spectral distribution 11, *12*, *13*, *14*
  wave–particle duality 10–11
light sources 11
  luminances 50
lightness 80
lightness constancy 80, 81–2
lightness contrast 81–2
long-sighted *see* hyperopia
luminance 79
  adaptation to changing conditions 50–57

coding for 79–82
interpretation of changes
    89–90
light sources 50
threshold for colour vision 51

M

Maxwell, James Clerk 10
medial superiortemporal area
    (cortex) *74*, 84
medial temporal area (cortex)
    74, 84
mental rotation 97
minutes of arc 61
modulation 63
**modulation transfer function
    (mtf) 62**–3, *64*
monochromatic sources **11**
motion parallax 84, 85–6
motion vision 7, 48, 84
mtf *see* modulation transfer
    function
myopia 31, 32, 53

N

near point 29
near reflex 22
Necker cube 99
neural adaptation 57
neural image 36, 41
neural nets 100–1

O

ocular dominance 85
OFF-centre cells **42**
ON-centre cells **42**, 76, 77
opsin 38, 39
optic chiasm 72, *73*, 74, 85
optic disc *21*
optic nerve *21*, *37*, 72, *73*
optic radiation 73
optic tract *73*
orientation-selective cells 78, 79

P

pandemonium model of
    perception **93**–4
**parallel distributed processing
    (PDP) 100**
parallel processing 95, 96, 99

PDP *see* parallel distributed
    processing
phobias 102
photons 10–11
photopic conditions **36**
photoreceptors 36–7, 41, 61–2,
    73
    see also cones; rods
plexiform layer *36*, *37*
polyene chain 39
Ponzo illusion 97, *98*
power of lens 23, **28**
presbyopia 32
primary colours 13
primary visual cortex 73–5
    see also V1 (cerebral cortex)
principle of superposition 24,
    34
prosopagnosia 96, 101
pupil of eye **21**, 22
    alterations in size 51–6
    see also direct light reflex
pursuit movements 48

Q

quantum theory 10

R

reading 7, 49, 94–5
receptive fields 41–3, 63, 76,
    77, 78, 80
    and binocular stereopsis 85
    and dark adaptation 57
    size variation over retina 43,
        68
recognition acuity 59
recognition of faces 95–7, 102
recognition of letters 93, 94–5,
    100, *101*
recognition of shapes 75, 86–91,
    93–102
'red eye' effect (photography)
    52
refractive power 29
refraction 23, 24–6
refractive index 25
relative reflectance 80–81
resolution acuity 59
    comparing 62
    definition 61
    estimation 59–61
    predicted value 61–2

retina 21, 36–43, 76
    anatomy 71–2
    dark adaptation 56–7
    neural adaptation in 57
    transduction in 36–7
11-*cis*-retinal 38, 39, 40
all-*trans*-retinal 38, 40
retinal ganglion cells (rgc) 36,
    37, **41**–3, 57, 68, 72, 76,
    80–81
retinotopic map 72, *73*, 74
rgc *see* retinal ganglion cells
rhodopsin 38, 39, 40, 57
rods (photoreceptors) **36**, *37*,
    42, 44–5
    distribution in retina 67
    photochemistry of vision in
        38–40, 57
    threshold of vision 51

S

saccades 46–7
saccadic omission 47
scotopic conditions **36**
    visual acuity in 67
segmentation 86
short-sighted *see* myopia
simple cells 78, 79, 80–81
size constancy 97, *98*
Snell's law of refraction 25, 26
spatial frequencies 15–18,
    63–5
spatial vision 76–9
spectacles 32
spectral distribution of light
    11, *12*, *13*, *14*
spectral reflectance 11
spherical aberrations 53, *54*
stereograms 86
stereopsis *see* binocular
    stereopsis
striate cortex 73
Stroop effect 7, 8
sunlight 9, *10*
superposition of waves *see*
    principle of superposition

T

temporal frequencies 15
texture 87–8
Thatcher illusion 97
Titchener circles 75

top-down processing **71**, 91, 99
transducin 40
trichromatic vision **82**

U

units
    power of lens 23, 28
    spatial frequencies 15

V

V1 (cerebral cortex) **73**, 74, 84,
    85
    cells in 76, 78–9
V4 (cerebral cortex) *74*, 82–3
value (colour science) **13**
vergence movements of eyes
    46
    see also disjunctive
        movements of eyes
vernier acuity 59
version movements of eyes **46**
vestibulo-ocular movements
    48–9
vestibulo-ocular reflex 48
visual acuity 58–68
visual angle 60
visual cortex, processing in
    76–9
visual illusions 90, 97, *98*
    Titchener circles 75
visual perception 5–6
visual sub-modalities 74,
    76–86
    coding of colour 82–3
    coding of luminance
        intensity 79–82
    coding of spatial structure
        76–9·
    measuring relative depth
        83–6
vitamin A 40

W

wavefronts 24–5
white light source **11**

Z

zonular fibres 21, 30